Geodetic and Control Surveying

US Army Corps of Engineers

University Press of the Pacific
Honolulu, Hawaii

Geodetic and Control Surveying

by
U.S. Army Corps of Engineers

ISBN: 1-4102-1835-X

DEPARTMENT OF THE ARMY
US Army Corps of Engineers
Washington, DC 20314-1000

EM 1110-1-1004

CECW-EE

Manual
No. 1110-1-1004

1 June 2002

Engineering and Design
GEODETIC AND CONTROL SURVEYING

1. Purpose. This manual provides technical specifications and procedural guidance for control and geodetic surveying. It is intended for use by engineering, topographic, and construction surveyors performing control surveys for civil works, military construction, and environmental restoration projects. Procedural and quality control standards are defined to establish uniformity in control survey performance and contract administration.

2. Applicability. This manual applies to all USACE commands having responsibility for the planning, engineering and design, operations, maintenance, construction, and related real estate and regulatory functions of civil works, military construction, and environmental restoration projects. It applies to control surveys performed by both hired-labor forces and contracted survey forces.

FOR THE COMMANDER:

4 Appendices
(See Table of Contents)

JOSEPH SCHROEDEL
Colonel, Corps of Engineers
Chief of Staff

DEPARTMENT OF THE ARMY
US Army Corps of Engineers
Washington, DC 20314-1000

EM 1110-1-1004

CECW-EE

Manual
No. 1110-1-1004

1 June 2002

Engineering and Design
GEODETIC AND CONTROL SURVEYING

Table of Contents

Subject	Paragraph	Page

Appendix A
References

Appendix B
CORPSCON Technical Documentation and Operating Instructions

Appendix C
Development and Implementation of NAVD 88

Appendix D
Requirements and Procedures for Referencing Coastal Navigation Projects to Mean Lower Low Water (MLLW) Datum

Glossary

Chapter 1
Introduction

1-1. Purpose

This manual provides technical specifications and procedural guidance for control and geodetic surveying. It is intended for use by engineering, topographic, and construction surveyors performing control surveys for civil works, military construction, and environmental restoration projects. Procedural and quality control standards are defined to establish uniformity in control survey performance and contract administration.

1-2. Applicability

This manual applies to all USACE commands having responsibility for the planning, engineering and design, operations, maintenance, construction, and related real estate and regulatory functions of civil works, military construction, and environmental restoration projects. It applies to control surveys performed by both hired-labor forces and contracted survey forces.

1-3. Distribution

This publication is approved for public release; distribution is unlimited.

1-4. References

Referenced USACE publications are listed in Appendix A. Where applicable, bibliographic information is listed within each chapter or appendix.

1-5. Background

A geodetic control survey consists of establishing the horizontal and vertical positions of points for the control of a project or installation site, map, GIS, or study area. These surveys establish three-dimensional point positions of fixed monuments, which then can provide the primary reference for subsequent engineering and construction projects. These control points also provide the basic framework from which detailed site plan topographic mapping, boundary demarcation, and construction alignment work can be performed. Precisely controlled monuments are also established to position marine construction vessels supporting the Corps navigation mission--e.g., the continuous positioning of dredges and survey boats. Geodetic control survey techniques are also used to effectively and efficiently monitor and evaluate external deformations in large structures, such as locks and dams.

1-6. Scope of Manual

This manual covers the use of engineering surveying techniques for establishing and/or extending project construction control. Accuracy requirements, standards, measurement procedures, calibrations, horizontal and vertical datum transformations, data reduction and adjustment methods, and engineering surveying techniques are outlined. The primary focus of this manual is on conventional (i.e., non-GPS) horizontal and vertical survey techniques using traditional ground survey instruments--transits, theodolites, levels, electronic total stations, etc. Typically, conventional survey techniques include traverse, triangulation, trilateration, and differential leveling.

a. The manual is intended to be a reference guide for control surveying, whether performed by in-house hired-labor forces, contracted forces, or combinations thereof. General planning criteria, field and office execution procedures, project datum requirements, and required accuracy specifications for performing engineering surveys are provided. Accuracy specifications, procedural criteria, and quality control requirements contained in this manual should be directly referenced in the scopes of work for Architect-Engineer (A-E) survey services or other third-party survey services. This ensures that standardized procedures are followed by both hired-labor and contract service sources.

b. The survey performance criteria given in this manual are not intended to meet the Federal Geodetic Control Subcommittee (FGCS) standards required for densifying the National Geodetic Reference System (NGRS). However, the methods and procedures given in this manual will yield results equal to or exceeding FGCS Second Order relative accuracy criteria. Second Order accuracy is generally considered sufficient for most USACE engineering and construction work.

c. This manual does not cover the concepts of using differential GPS for performing precise geodetic control surveys. For further specific guidance on all aspects of GPS surveying, the user should consult EM 1110-1-1003, NAVSTAR GPS Surveying.

d. This manual should be used in conjunction with other USACE surveying and mapping engineering manuals that refer to it for guidance on datums and datum transformation procedures. These procedures are covered in Chapter 4 (Reference Systems and Datum Transformations) and in Appendices B, C, and D.

e. This manual was initially developed as part of the 31 October 1994 version of EM 1110-1-1004, "Deformation and Control Surveying." During the current update, structural deformation surveying portions of the 1994 manual were removed and incorporated into a separate technical manual. The current version of EM 1110-1-1004 was then re-titled as "Geodetic and Control Surveying" to reflect the revised scope.

1-7. Life Cycle Project Management

Project control surveys may be required through the entire life cycle of a project, spanning decades in many cases. During the early planning phases of a project, a comprehensive control plan should be developed which considers survey requirements over a project's life cycle, with a goal of eliminating duplicate or redundant surveys to the maximum extent possible.

1-8. Metrics

Both English and metric units are used in this manual. Metric units are commonly used in precise surveying applications, including the horizontal and vertical survey work covered in this manual. Control survey measurements are usually recorded and reported in metric units. In all cases, the use of either metric or English units shall follow local engineering and construction practices.

1-9. Trade Name Exclusions

The citation or illustration in this manual of trade names of commercially available survey products, including other auxiliary surveying equipment, instrumentation, and adjustment software, does not constitute official endorsement or approval of the use of such products.

1-10. Abbreviations and Terms

Abbreviations used in this manual are defined in the Glossary at the end of this manual. Commonly used engineering surveying terms are also explained in the Glossary.

1-11. Mandatory Requirements

ER 1110-2-1150 (Engineering and Design for Civil Works Projects) prescribes that mandatory requirements be identified in engineer manuals. Mandatory requirements in this manual are summarized at the end of each chapter. Mandatory accuracy standards, quality control, and quality assurance criteria are normally summarized in tables within each chapter. The mandatory criteria contained in this manual are based on the following considerations: (1) project safety considerations, (2) overall project function, (3) previous Corps experience and practice has demonstrated the criteria are critical, (4) Corps-wide geospatial data standardization requirements, (5) adverse economic impacts if criteria are not followed, and (6) HQUSACE commitments to Federal and industry standards.

1-12. Proponency

The HQUSACE proponent for this manual is the Engineering and Construction Division, Directorate of Civil Works (CECW-EE). Technical development and compilation of the manual was coordinated by the US Army Topographic Engineering Center (CEERD-TS-G). Comments, recommended changes, or waivers to this manual should be forwarded through MSC to HQUSACE (ATTN: CECW-EE).

Chapter 2
Control Surveying Applications

2-1. General

Control surveys are used to support a variety of USACE project applications. These include project boundary control densification, structural deformation studies, photogrammetric mapping, dynamic positioning and navigation for hydrographic survey vessels and dredges, hydraulic study/survey location, river/floodplain cross-section location, core drilling location, environmental studies, levee overbank surveys, levee profiling, levee grading and revetment placement, disposal area construction, grade control, real property surveys, and regulatory enforcement actions. Some of these applications are described below.

2-2. Project Control Densification

a. Conventional surveying. Conventional geodetic control surveys are those performed using traditional precise surveying techniques and instruments--i.e., theodolites, total stations, and levels. Conventional control surveys can be used to economically and accurately establish or densify project control in a timely fashion. Quality control statistics and redundant measurements in networks established by these methods help to ensure reliable results. However, conventional survey methods do have the requirement for intervisibility between adjacent stations.

b. GPS surveying. GPS satellite surveying techniques can often be used to establish or densify project control more efficiently (and accurately) than conventional control surveying techniques--especially over large projects. As with conventional methods, quality control statistics and redundant measurements in GPS networks help to ensure reliable results. Field operations to perform a GPS survey are relatively easy and can generally be performed by one person per receiver, with two or more receivers required to transfer control. GPS does not require intervisibility between adjacent stations. However, GPS must have visibility of at least four satellites during surveying. This requirement may make GPS inappropriate in areas of dense vegetation. For GPS control survey techniques refer to EM 1110-1-1003, NAVSTAR GPS Surveying.

2-3. Geodetic Control Densification

Conventional control and GPS surveying methods can be used for wide-area, high-order geodetic control densification. First-, Second- or Third-Order work can be achieved using conventional or GPS surveying techniques. GPS techniques are now generally used for most horizontal control surveys performed for mapping frameworks. Conventional instruments and procedures are generally preferred for site plan topographic mapping and critical construction control. Topographic mapping procedures used in detailed site plan surveys are contained in EM 1110-1-1005, Topographic Surveying.

2-4. Vertical Control Densification

Conventional leveling methods are used to determine orthometric height elevations of benchmarks established for vertical control densification. The setup and operation for conventional control surveying for vertical control densification offers economies of scale in the same manner as that offered by the setup for horizontal project control densification--i.e., smaller projects require less setups, while larger projects require more. For large mapping projects, differential GPS may prove more cost effective for densifying vertical control. However, for small project sites or construction projects, conventional spirit leveling is generally preferred.

2-5. Structural Deformation Studies

a. Conventional control surveying can be used to monitor the motion of points on a structure relative to stable monuments. This is usually done using an Electronic Distance Measuring (EDM) instrument located on various stable reference monuments away from the structure, and measuring precise distances to calibrated reflectors positioned at selected points on the structure. When only distances are measured, trilateration techniques may be employed to compute absolute movements. If angular observations are added, such as with a theodolite or electronic total station, then triangulation methods may be added to a position solution. These precise techniques can provide a direct measure of the displacement of a structure as a function of time. If procedures are strictly adhered to, it is possible to achieve a ± 0.5 mm + 4 ppm (4 mm/km) baseline accuracy using conventional surveying instruments. Personnel requirements generally are two, once the initial test network of reference and object points are set up--one person to monitor the EDM or total station and another to aid in reflector placement.

b. GPS can also be used to monitor the motion of points on a structure relative to stable monuments. With GPS, an array of antennae are positioned at selected points on the structure and on remote stable monuments--as opposed to using reflectors and EDMs as previously described. The baselines between the antennae are formulated to monitor differential movement. The relative precision of the measurements is on the order of ± 5 mm over distances averaging between 5 and 10 km, and near the 1-mm level for short baselines. GPS observations can be determined continuously 24 hours a day. Once a deformation monitoring system has been set up using GPS, it can be operated unattended and is relatively easy to maintain.

2-6. Photogrammetry

Geodetic control surveys are used in the support of photogrammetric mapping applications. These control surveys are performed to provide rigid horizontal and vertical alignment of the photographs. Since photogrammetric mapping projects typically are large in extent, GPS methods have largely replaced conventional control survey techniques. In many cases, photogrammetric mapping control surveys have been largely eliminated through the use of differential GPS-controlled airborne cameras. More specific guidance on the use of control surveying in support of photogrammetry is included in EM 1110-1-1000, Photogrammetric Mapping.

2-7. Dynamic Positioning and Navigation

a. Conventional control surveying can be used to establish the primary project control for the dynamic positioning and navigation of construction and surveying platforms used for design, construction, and environmental regulatory efforts. These efforts include dredge control systems, site investigation studies/surveys, horizontal and vertical construction placement, hydraulic studies, or any other waterborne activity requiring two- or three-dimensional control. Second Order or Third Order leveling is required for these efforts.

b. GPS has reduced (or even eliminated in many cases) the time and effort required to establish control for dynamic positioning and navigation systems. In addition to this capability, GPS equipment can provide dynamic, real-time GPS code and carrier phase positioning of construction and surveying platforms. GPS code phase differential techniques can provide real-time meter-level horizontal positioning and navigation, while GPS carrier phase differential techniques can provide real-time, centimeter-level, three-dimensional positioning and navigation. These GPS methods can be used for any type of construction or survey platform (e.g., dredges, graders, survey vessels, etc.). More specific guidance on the use of GPS for dynamic positioning and navigation is included in EM 1110-2-1003, Hydrographic Surveying.

2-8. GIS Integration

A Geographic Information System (GIS) can be used to correlate and store diverse information on natural or man-made characteristics of geographic positions. To effectively establish and use a GIS, it must be based on accurate geographic coordinates. A GIS with an accurate foundation of geographic coordinates enables the user to readily exchange information between databases. Conventional control surveying and GPS surveying can be used to establish the geographic coordinates used as the foundation for a GIS. Refer to EM 1110-1-2909, Geospatial Data and Systems, for detailed guidance on GIS development.

Chapter 3
Standards and Specifications for Control Surveying

3-1. General

This chapter details standards and specifications for control surveying and provides guidance on how to achieve them.

3-2. Accuracy

The accuracy of control surveying measurements should be consistent with the purpose of the survey. When evaluating the techniques to be used and accuracies desired, the surveyor must evaluate the limits of the errors of the equipment involved, the procedures to be followed, and the error propagation. These evaluations should be firmly based on past experience or written guidance. It is important to remember in this evaluation that the best survey is the one that provides the data at the required accuracy levels, without wasting manpower, time, and money.

   a. Survey accuracy standards prescribed in this section relate to the relative accuracy derived from a particular survey. This relative accuracy (or precision) is estimated by internal closure checks of the survey run through the local project, map, or construction site. Relative survey accuracy estimates are traditionally expressed as ratios of the misclosure to the total length of the survey (e.g., 1:10,000). Relative survey accuracies are different than map accuracies, which are expressed in terms of limiting positional error. Since map compilation is dependent on survey control, map accuracies will ultimately hinge on the adequacy and accuracy of the base survey used to control the map.

   b. Tables 3-1 and 3-2 detail the basic minimum criteria required for planning, performing, and evaluating the adequacy of control surveys.

   c. These criteria apply to all conventional control surveying as well as GPS surveying activities-- the intended accuracy is independent of the survey equipment employed.

   d. Many of the criteria shown in the tables are developed from FGCS standards for performing conventional control surveys and GPS surveys. The criteria listed in the tables have been modified to provide more practical standards for engineering and survey densification. FGCS Standards and Specifications for Geodetic Control Networks (FGCS 1984) covers all aspects of performing conventional control surveys for high-precision geodetic network densification purposes, while FGCS GPS survey standards (FGCS 1988) covers the use of GPS surveys for the same application. The application of some FGDC standards and techniques are not always practical for typical civil works, military construction, and environmental restoration activities where lower precision control is acceptable.

   e. If a primary function of a survey is to support NGRS densification, then specifications listed in the FGCS publications (FGCS 1984 and FGCS 1988) should be followed in lieu of those in Tables 3-1 and 3-2.

**Table 3-1. USACE Point Closure Standards
for Horizontal Control Surveys (ratio)**

USACE Classification	Point Closure Standard
Second Order Class I	1:50,000
Second Order Class II	1:20,000
Third Order Class I	1:10,000
Third Order Class II	1: 5,000
Fourth Order	1:2,500 - 1:20,000

**Table 3-2. USACE Point Closure Standards
for Vertical Control Surveys (in feet)**

USACE Classification	Point Closure Standard
Second Order Class I	0.025*sqrt M
Second Order Class II	0.035*sqrt M
Third Order	0.050*sqrt M
Fourth Order	0.100*sqrt M

where sqrt M = square root of distance M in miles

(1) Survey classification. A survey shall be classified based on its horizontal point closure ratio, as indicated in Table 3-1, or the vertical elevation difference closure standard given in Table 3-2.

(2) Horizontal control standards. The horizontal point closure is determined by dividing the linear distance misclosure of the survey into the overall circuit length of a traverse, loop, or network line/circuit. When independent directions or angles are observed, as on a conventional survey (i.e., traverse, trilateration, or triangulation), these angular misclosures may optionally be distributed before assessing positional misclosure. In cases where GPS vectors are measured in geocentric coordinates, then the three-dimensional positional misclosure is assessed.

(a) Approximate surveying. Approximate surveying work should be classified based on the survey's estimated or observed positional errors. This would include absolute GPS and some differential GPS techniques with positional accuracies ranging from 10 to 150 feet (95% RMS). There is no order classification for such approximate work.

(b) Higher order survey. Requirements for relative line accuracies exceeding 1:50,000 are rare for most USACE applications. Surveys requiring accuracies of First Order (1:100,000) or better should be performed using FGCS standards and specifications, and must be adjusted by the NGS.

(c) Construction layout or grade control. This classification is analogous to traditional Fourth Order work. It is intended to cover temporary control used for alignment, grading, and measurement of various types of construction, and some local site plan topographic mapping or photo mapping control work. Accuracy standards will vary with the type of construction. Lower accuracies (1:2,500 - 1:5,000) are acceptable for earthwork, embankment, beach fill, and levee alignment stakeout and grading, and some site plan, curb and gutter, utility building foundation, sidewalk, and small roadway stakeout. Moderate accuracies (1:5,000) are used in most pipeline, sewer, culvert, catch basin, and manhole stakeout and for general residential building foundation and footing construction, major highway pavement, and concrete runway stakeout work. Somewhat higher accuracies (1:10,000 -1: 20,000) are used for aligning longer bridge spans, tunnels, and large commercial structures. For extensive bridge or tunnel projects, 1:50,000 or even 1:100,000 relative accuracy alignment work may be required. Vertical grade is usually observed to the nearest 0.01 foot for most construction work, although 0.1-foot accuracy is sufficient for riprap placement, grading, and small diameter pipe placement. Construction control points are typically marked by semi-permanent or temporary monuments (e.g., plastic hubs, P-K nails, wooden grade stakes). Control may be established by short, non-redundant spur shots, using total stations or GPS, or by single traverse runs between two existing permanent control points. Positional accuracy will be commensurate with, and relative to, that of the existing point(s) from which the new point is established.

(3) Vertical control standards. The vertical accuracy of a survey is determined by the elevation misclosure within a level section or level loop. For conventional differential or trigonometric leveling, section or loop misclosures (in feet) shall not exceed the limits shown in Table 3-2, where the line or circuit length (M) is measured in miles. Fourth Order accuracies are intended for construction layout grading work. Procedural specifications or restrictions pertaining to vertical control surveying methods or equipment should not be overly restrictive.

(4) Contract compliance with USACE survey standards. Contract compliance assessment shall be based on the prescribed point closure standards of internal loops, not on closure with external networks of unknown accuracy. In cases where internal loops are not observed, then assessment must be based on external closures. Specified closure accuracy standards shall not be specified that exceed those required for the project, regardless of the accuracy capabilities of the survey equipment.

3-3. General Procedural Standards and Specifications

a. Most survey applications in typical civil works and military arenas can be satisfied with a Second- or Third Order level of accuracy. Higher levels of accuracy are required for the densification of high-precision geodetic networks and some forms of deformation monitoring.

b. Since most modern survey equipment (e.g., GPS or electronic total stations) are capable of achieving far higher accuracies than those required for engineering, construction, and mapping, only generalized field survey specifications are necessary for most USACE work. The following paragraphs outline some of the more critical specifications that relate to the USACE horizontal and vertical standards. Additional guidance for performing control surveying is found in subsequent chapters, as well as in some of the technical manuals listed in Appendix A.

(1) Survey instrumentation criteria. USACE Commands shall minimize the use of rigid requirements for particular surveying equipment or instruments used by professional surveying contractors. In some instances, contract technical specifications may prescribe a general type of instrument system be employed (e.g., total station, GPS, spirit level), along with any unique operating or calibration requirements.

(2) Survey geometry and field observing criteria. In lieu of providing detailed government procedural specifications, professional contractors may be presumed capable of performing surveys in accordance with accepted industry standards and practices. Any geometrical form of survey network may be formed: traverses, loops, networks, and cross-links within networks. Traverses should generally be closed back (or looped) to the same point, to allow an assessment of the internal misclosure accuracies. Survey alignment, orientation, and observing criteria should not be rigidly specified; however, guidance regarding limits on numbers of traverse stations, minimum traverse course lengths, auxiliary connections, etc. are provided in the subsequent chapters of this EM, as well in other EMs listed in Appendix A.

(3) Connections to existing control. USACE surveys should be connected to existing local control or project control monuments/benchmarks. These existing points may be those of any Federal (including USACE project control), state, local, or private agency. Ties to local USACE project control and boundary monuments are absolutely essential and critical to design, construction, and real estate. In order to minimize scale or orientation errors, at least two existing monuments should be connected, if practicable. However, survey quality control accuracy assessments (Tables 3-1 and 3-2) shall be based on internal traverse or level line closures--not on external closures between or with existing monuments or benchmarks. Accuracy assessments based on external closures typically require a knowledge of the statistical variances in the fixed network.

(4) Connections to NGRS control. The NGRS pertains to geodetic control monuments with coordinate or elevation data published by the NGS. It is recommended that USACE surveys be connected with one or more stations on the NGRS when practicable and feasible. Connections with the NGRS shall be subordinate to the requirements for connections with local/project control. Connections with local/project control that has previously been connected to the NGRS are adequate in most cases.

(5) Survey datums. A variety of survey datums and references are used throughout USACE projects. It is recommended that horizontal surveys be referenced to the North American Datum of 1983 (NAD 83) or, if NAD 83 is unavailable, to the NAD 27 system, with coordinates referred to the local State Plane Coordinate System (SPCS) for the area. The NAD 83 is the preferred reference datum. The Universal Transverse Mercator (UTM) grid system may be used for military operational or tactical uses, in OCONUS locales without a local coordinate system, or on some civil projects crossing multiple SPCS zones. Vertical control should be referenced to either the National Geodetic Vertical Datum, 1929 Adjustment (NGVD 29) or the North American Vertical Datum, 1988 Adjustment (NAVD 88). Independent survey datums and reference systems shall be avoided unless required by local code, statute, or practice. This includes local tangent grid systems, state High Accuracy Reference Networks (HARN), and unreferenced construction baseline station-offset control.

(6) Spur points. Spur points (open-ended traverses or level lines) should be avoided to the maximum extent practicable. In many cases, it is acceptable survey procedure for temporary Fourth Order construction control, provided adequate blunder detection is taken. Kinematic differential GPS (DGPS code or carrier phase tracking) surveys are effectively spur point surveys with relative accuracies well in excess of Third Order standards. These DGPS kinematic spur techniques may be acceptable procedures for most control surveying in the future, provided blunder protection procedures are developed. Refer to EM 1110-1-1003 for further GPS guidance.

(7) Survey adjustments. The standard adjustment method in USACE will be either the Compass Rule or Least Squares. Technical (and contractual) compliance with accuracy standards prescribed in Tables 3-1 and 3-2 will be based on the internal point misclosures. Propagated relative distance/line accuracy statistics used by FGCS that result from unconstrained (minimally constrained) least squares adjustment error propagation statistics may be assumed comparable to relative misclosure accuracy estimates for survey quality control assessment. Surveys may be adjusted (i.e., constrained) to existing control without regard to the variances in the existing network adjustment. Exceptions to this requirement are surveys performed to FGCS standards and specifications that are adjusted by NGS.

(8) Data recording and archiving. Field survey data may be recorded either manually or electronically. Manual recordation should follow industry practice, using formats outlined in various technical manuals (Appendix A). Refer to EM 1110-1-1005, Topographic Surveying, for electronic survey data collection standards.

3-4. Construction Surveys

In-house and contracted construction surveys generally will be performed to meet Third Order-Class II (1:5,000) accuracy. Some stakeout work for earthwork clearing and grading, and other purposes, may need only be performed to meet Fourth Order accuracy requirements. Other stakeout work, such as tunnel or bridge pier alignment, may require Second Order or higher accuracy criteria. Construction survey procedural specifications should follow recognized industry practices.

3-5. Cadastral and Real Estate Surveys

Many state codes, rules, or statutes prescribe minimum technical standards for surveying and mapping. Generally, most state accuracy standards for real property surveys parallel USACE Third Order point closure standards--usually ranging between 1:5,000 and 1:10,000. USACE and its contractors shall follow applicable state minimum technical standards for real property surveys involving the determination of the perimeters of a parcel or tract of land by establishing or reestablishing corners, monuments, and boundary lines, for the purpose of describing, locating fixed improvements, or platting or dividing parcels. Although state minimum standards relate primarily to accuracies of land and boundary surveys, other types of survey work may also be covered in some areas. See also ER 405-1-12, Real Estate Handbook. Reference also the standards and specifications prescribed in the "Manual of Instruction for the Survey of the Public Lands of the United States" (US Bureau of Land Management 1947) for cadastral surveys, or surveys of private lands abutting or adjoining Government lands.

3-6. Geodetic Control Surveys

a. Geodetic control surveys are usually performed for the purpose of establishing a basic framework to be included in the national geodetic reference network, or NGRS. These geodetic survey functions are distinct from the survey procedures and standards defined in this EM which are intended to support USACE engineering, construction, mapping, and Geographic Information System (GIS) activities.

b. Geodetic control surveys of permanently monumented control points that are to be incorporated in the NGRS must be performed to far more rigorous standards and specifications than are surveys used for general engineering, construction, mapping, or cadastral purposes. When a project requires NGRS densification, or such densification is a desirable by-product and is economically justified, USACE Commands should conform to FGCS survey standards and specifications, and other criteria prescribed under Office of Management and Budget (OMB) Circular A-16 (OMB 1990). This includes related automated data recording, submittal, project review, and adjustment requirements mandated by FGCS and NGS. Details outlining the proposed use of FGCS standards and specifications in lieu of USACE standards, including specific requirements for connections to the NGRS, shall be included in the descriptions of survey and mapping activities contained in project authorization documents.

c. Geodetic control surveys intended for support to and inclusion in the NGRS must be done in accordance with the following FGCS publications:

(1) "Standards and Specifications for Geodetic Control Networks" (FGCS 1984).

(2) "Geometric Geodetic Accuracy Standards and Specifications for Using GPS Relative Positioning Techniques" (FGCS 1988).

(3) "Input Formats and Specifications of the National Geodetic Data Base" (also termed the "Bluebook") (FGCS 1980).

(4) "Guidelines for Submitting GPS Relative Positioning Data to the National Geodetic Survey" (NGS 1988).

A survey performed to FGCS accuracy standards and specifications cannot be definitively classified or certified until the NGS has performed a variance analysis of the survey relative to the existing NGRS. This analysis and certification cannot be performed by USACE Commands-- only the NGS can perform this function. It is estimated that performing surveys to meet these FGCS standards, specifications, and archiving criteria can add between 25 and 50 percent to the surveying costs of a project. Therefore, sound judgment must be exercised on each project when determining the practicability of doing survey work that, in addition to meeting the needs of the project, can be used for support to and inclusion in the NGRS.

3-7. Topographic Site Plan Mapping Surveys

Control surveys from which site plan mapping is densified (using plane tables, electronic total stations, or GPS) are normally established to USACE Third Order standards. Follow guidance in EM 1110-1-1005, Topographic Surveying.

3-8. Structural Deformation Surveys

Structural deformation surveys are performed in compliance with the requirements in ER 1110-2-100, often termed PICES surveys. PICES surveys require high line vector and/or positional accuracies to monitor the relative movement of monoliths, walls, embankments, etc. PICES survey accuracy standards vary with the type of construction, structural stability, failure probability, and impact, etc. In general, horizontal and vertical deformation monitoring survey procedures are performed relative to a control network established for the structure. Ties to the NGRS or NGVD 29 are not necessary other than for general reference; and then only an USACE Third Order connection is needed. FGCS geodetic relative accuracy standards are not applicable to these localized movement surveys. Other deformation survey and instrumentation specifications and procedures for earth and rock fill dams and concrete structures are in EM 1110-2-1908 and EM 1110-2-4300.

3-9. Photogrammetric Mapping Control Surveys

Control surveys required for controlling photogrammetric mapping products will normally be performed to USACE Third Order standards. Occasionally USACE Second Order standards will be required for extensive aerotriangulation work. Reference EM 1110-1-1000, Photogrammetric Mapping, for detailed photogrammetric mapping standards and specifications.

3-10. Hydrographic Surveys

Control points for USACE hydrographic surveys generally are set to Third Order horizontal and vertical accuracy. Exceptions are noted in EM 1110-2-1003, Hydrographic Surveying. Hydrographic depth sounding accuracies are based on the linear and radial error measures described in EM 1110-2-1003, as well as hydrographic survey procedural specifications.

3-11. GIS Surveys

GIS raster or vector features can be scaled or digitized from any existing map. Typically a standard USGS 1:24,000 quadrangle map is adequate given the accuracies needed between GIS data features, elements, or classifications. Relative or absolute GPS (i.e., 10- to 30-foot precision) survey techniques may be adequate to tie GIS features where no maps exist. Second- or Third Order control networks are generally adequate for all subsequent engineering, construction, real estate, GIS, and/or Automated Mapping/Facilities Management (AM/FM) control.

3-12. Mandatory Requirements

The accuracy standards in Tables 3-1 and 3-2 are mandatory.

Chapter 4
Reference Systems and Datum Transformations

4-1. Reference Systems

a. General. The discipline of surveying consists of locating points of interest on the surface of the earth. The positions of points of interest are defined by coordinate values that are referenced to a predefined mathematical surface. In geodetic surveying, this mathematical surface is called a datum, and the position of a point with respect to the datum is defined by its coordinates. The reference surface for a system of control points is specified by its position with respect to the earth and its size and shape. A datum is a coordinate surface used as reference figure for positioning control points. Control points are points with known relative positions tied together in a network. Densification of the network of control points refers to adding more control points to the network and increasing its scope. Both horizontal and vertical datums are commonly used in surveying and mapping to reference coordinates of points in a network. Reference systems can be based on the geoid, ellipsoid, or a plane. The physical earth's gravity force can be modeled to create a positioning reference frame that rotates with the earth. The geoid is such a surface (an equipotential surface of the earth's gravity field) that best approximates Mean Sea Level (MSL)--see Figure 4-1. The orientation of this surface at a given point on geoid is defined by the plumb line. The plumb line is oriented tangent to the local gravity vector. Surveying instruments can be readily oriented with respect to the gravity field because its physical forces can be sensed with simple mechanical devices. A mean gravity field can be used as a reference surface to represent the actual earth's gravity field. Such a reference surface is developed from an ellipsoid of revolution that best approximates the geoid. An ellipsoid of revolution provides a well-defined mathematical surface to calculate geodetic distances, azimuths, and coordinates. The major semi-axis (a) and minor semi-axis (b) are the parameters used to determine the ellipsoid size and shape. The shape of a reference ellipsoid also can be described by either its flattening (f) or its eccentricity (e).

Flattening: $\qquad f = (a - b) / a$

Eccentricity: $\qquad e = [sqrt(a^2 - b^2)] / a$

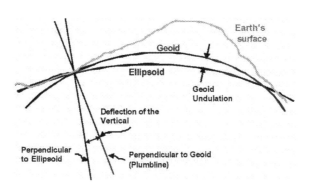

Figure 4-1. The relationship between the ellipsoid, geoid, and the physical surface of the earth

4-2. Geodetic Coordinates

a. *General.* A coordinate system is defined by the location of the origin, orientation of its axes, and the parameters (coordinate components) which define the position of a point within the coordinate system. Terrestrial coordinate systems are widely used to define the position of points on the terrain because they are fixed to the earth and rotate with it. The origin of terrestrial systems can be specified as either geocentric (origin at the center of the earth) or topocentric (origin at a point on the surface of the earth). The orientation of terrestrial coordinate systems is described with respect to its poles, planes, and axes. The primary pole is the axis of symmetry of the coordinate system, usually parallel to the rotation axis of the earth, and coincident with the minor semi-axis of the reference ellipsoid. The reference planes that are perpendicular to the primary pole are the equator (zero latitude) and the Greenwich meridian plane (zero longitude). Parameters for point positioning within a coordinate system refer to the coordinate components of the system (either Cartesian or curvilinear).

b. *Geodetic Coordinates.* Geodetic coordinate components consist of:

- latitude (ϕ),
- longitude (λ),
- ellipsoid height (h).

Geodetic latitude, longitude, and ellipsoid height define the position of a point on the surface of the Earth with respect to the reference ellipsoid.

(1) Geodetic latitude (ϕ). The geodetic latitude of a point is the acute angular distance between the equatorial plane and the normal through the point on the ellipsoid measured in the meridian plane (Figure 4-2). Geodetic latitude is positive north of the equator and negative south of the equator.

(2) Geodetic longitude (λ). The geodetic longitude is the angle measured counter-clockwise (east), in the equatorial plane, starting from the prime meridian (Greenwich meridian), to the meridian of the defined point (Figure 4-2). In the continental United States, longitude is commonly reported as a west longitude. To convert easterly to westerly referenced longitudes, the easterly longitude must be subtracted from 360 deg.

East-West Longitude Conversion:

$$\lambda\,(W) = [\,360 - \lambda\,(E)\,]$$

For example:

$$\lambda\,(E) = 282^{\,d}\,52^{\,m}\,36.345^{\,s}\,E$$
$$\lambda\,(W) = [\,360^{\,d} - 282^{\,d}\,52^{\,m}\,36.345^{\,s}\,E\,]$$
$$\lambda\,(W) = 77^{\,d}\,07^{\,m}\,23.655^{\,s}\,W$$

(3) Ellipsoid Height (h). The ellipsoid height is the linear distance above the reference ellipsoid measured along the ellipsoidal normal to the point in question. The ellipsoid height is positive if the reference ellipsoid is below the topographic surface and negative if the ellipsoid is above the topographic surface.

(4) Geoid Separation (N). The geoid separation (geoidal height) is the distance between the reference ellipsoid surface and the geoid surface measured along the ellipsoid normal. The geoid separation is positive if the geoid is above the ellipsoid and negative if the geoid is below the ellipsoid.

(5) Orthometric Height (H). The orthometric height is the vertical distance of a point above or below the geoid.

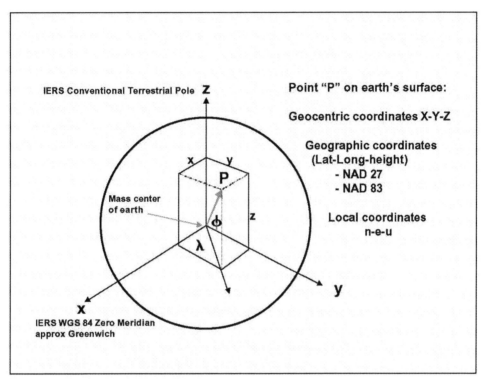

Figure 4-2. Coordinate reference frames

c. *Datums.* A datum is a coordinate surface used as reference for positioning control points. Both horizontal and vertical datums are commonly used in surveying and mapping to reference coordinates of points in a network.

(1) Horizontal Datum. A horizontal datum is defined by specifying: the 2D geometric surface (plane, ellipsoid, sphere) used in coordinate, distance, and directional calculations; the initial reference point (origin); and a defined orientation, azimuth or bearing from the initial point.

(a) Geodetic Datum. Five parameters are required to define an ellipsoid-based datum. The major semi-axis (a) and flattening (f) define the size and shape of the reference ellipsoid; the latitude and longitude of an initial point; and a defined azimuth from the initial point define its orientation with respect to the earth. The NAD 27 and NAD 83 systems are examples of horizontal geodetic datums.

(b) Project Datum. A project datum is defined relative to local control and might not be directly referenced to a geodetic datum. Project datums are usually defined by a system with perpendicular axes, and with arbitrary coordinates for the initial point, and with one (principal) axis oriented toward true north.

(d) NAD 27. NAD 27 is based on an adjustment of surveying measurements made between numerous control points using the Clarke 1866 reference ellipsoid. The origin and orientation of NAD 27 is defined relative to a fixed triangulation station in Kansas (i.e., Meades Ranch). Azimuth orientation for NAD 27 is referenced to South, with the Greenwich Meridian for longitude origin. The distance reference units for NAD 27 are in US Survey Feet. NAD 27 was selected for North America.

(e) NAD 83. NAD 83 is defined with respect to the Geodetic Reference System of 1980 (GRS 80) ellipsoid. GRS 80 is a geocentric reference ellipsoid. Azimuth orientation for NAD 83 is referenced to North with the Greenwich Meridian for longitude origin. The distance reference units for NAD 83 are in meters.

(f) WGS 84. WGS 84 is defined with respect to the World Geodetic System of 1984 (WGS 84) ellipsoid. WGS 84 is a geocentric reference ellipsoid and is the reference system for the Global Positioning System (GPS). Azimuth orientation for WGS 84 is referenced to North, with the Greenwich Meridian for longitude origin. The distance reference units for WGS 84 are in meters.

(2) Vertical Datum. A vertical datum is a reference system used for reporting elevations. Vertical datums are most commonly referenced to:

- Mean Sea Level (MSL),
- Mean Low Water (MLW),
- Mean Lower Low Water (MLLW),
- Mean High Water (MHW).

Mean Sea Level based elevations are used for most construction, photogrammetric, geodetic, and topographic surveys. MLLW elevations are used in referencing coastal navigation projects. MHW elevations are used in construction projects involving bridges over navigable waterways.

(a) The vertical reference system formerly used by USACE was the National Geodetic Vertical Datum of 1929 (NGVD 29). The North American Vertical Datum of 1988 (NAVD 88) should be used by USACE for all vertical positioning surveying work. Transformations between NGVD 29 and NAVD 88 have been developed for general use--refer to Appendix C for details.

4-3. State Plane Coordinate System

a. General. State Plane Coordinate Systems (SPCS) were developed by the National Geodetic Survey (NGS) to provide plane coordinates over a limited region of the earth's surface. To properly relate geodetic coordinates (ϕ-λ-h) of a point to a 2D plane coordinate representation (Northing, Easting), a conformal mapping projection must be used. Conformal projections have mathematical properties that preserve differentially small shapes and angular relationships as a result of the transformation between the ellipsoid and mapping plane. Map projections that are most commonly used for large regions are based on either a conic or a cylindrical mapping surface (Figure 4-3). The projection of choice is dependent on the north-south or east-west areal extent of the region. Areas with limited east-west dimensions and indefinite north-south extent use the Transverse Mercator (TM) type projection. Areas with limited north-south dimensions and indefinite east-west extent use the Lambert projection. The SPCS was designed to minimize the spatial distortion at a given point to approximately one part in ten thousand (1:10,000). To satisfy this criteria, the SPCS has been divided into zones that have a maximum width or height of approximately one hundred and fifty eight statute miles (158 miles). Therefore, each state may have several zones or may employ both the Lambert (conic) and Transverse Mercator (cylindrical) projections. The projection state plane coordinates must be referenced to a specific geodetic datum (i.e., the datum that the initial geodetic coordinates are referenced to must be known).

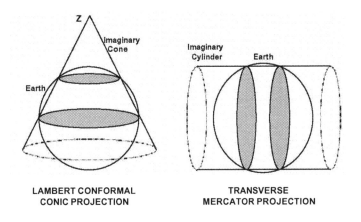

Figure 4-3. Common Map Projections

b. Transverse Mercator (TM). The Transverse Mercator projection uses a cylindrical surface to cover limited zones on either side of a central reference longitude. Its primary axis is rotated perpendicular to the symmetry axis of the reference ellipsoid. Thus, the TM projection surface intersects the ellipsoid along two lines equidistant from the designated central meridian longitude (Figure 4-4). Distortions in the TM projection increase predominantly in the east-west direction. The scale factor for the Transverse Mercator projection is unity where the cylinder intersects the ellipsoid. The scale factor is less than one between the lines of intersection, and greater than one outside the lines of intersection. The scale factor is the ratio of arc length on the projection to arc length on the ellipsoid. To compute the state plane coordinates of a point, the latitude and longitude of the point and the projection parameters for a particular TM zone or state must be known.

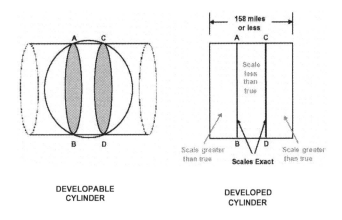

Figure 4-4. Transverse Mercator Projection

c. *Lambert Conformal Conic (LCC)*. The Lambert projection uses a conic surface to cover limited zones of latitude adjacent to two parallels of latitude. Its primary axis is coincident with the symmetry axis of the reference ellipsoid. Thus, the LCC projection intersects the ellipsoid along two standard parallels (Figure 4-5). Distortions in the LCC projection increase predominantly in the north-south direction. The scale factor for the Lambert projection is equal to unity at each standard parallel and is less than one inside, and greater than one outside the standard parallels. The scale factor remains constant along the standard parallels.

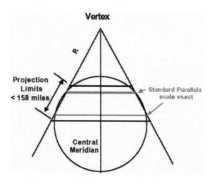

Figure 4-5. Lambert Projection

c. *Scale units*. State plane coordinates can be expressed in both feet and meters. State plane coordinates defined on the NAD 27 datum are published in feet. State plane coordinates defined on the NAD 83 datum are published in meters, however, state and federal agencies can request the NGS to provide coordinates in feet. If NAD 83 based state plane coordinates are defined in meters and the user intends to convert those values to feet, the proper meter-feet conversion factor must be used. Some states use the International survey foot rather than the US Survey foot in the conversion of feet to meters.

International Survey Foot:

  1 International Foot = 0.3048 meter (exact)

US Survey Foot:

  1 US Survey Foot = 1200 / 3937 meter (exact)

4-4. Universal Transverse Mercator Coordinate System

a. *General*. Universal Transverse Mercator (UTM) coordinates are used in surveying and mapping when the size of the project extends through several state plane zones or projections. UTM coordinates are also utilized by the US Army, Air Force, and Navy for mapping, charting, and geodetic applications. The UTM projection differs from the TM projection in the scale at the central meridian, origin, and unit representation. The scale at the central meridian of the UTM projection is 0.9996. In the Northern Hemisphere, the northing coordinate has an origin of zero at the equator. In the Southern Hemisphere, the southing coordinate has an origin of ten million meters (10,000,000 m). The easting coordinate has an origin five hundred thousand meters (500,000 m) at the central meridian. The UTM system is divided into sixty (60) longitudinal zones. Each zone is six (6) degrees in width extending three (3) degrees on each side of the central meridian. The UTM system is applicable between latitudes eighty-four degrees north (84 ^{d} N) to eighty degrees south (80 ^{d} S). To compute the UTM coordinates of a point, the TM coordinates must be determined. The UTM northing or southing (N $_{UTM}$, S $_{UTM}$) coordinates are computed by multiplying the scale factor (0.9996) at the central meridian by the TM northing or southing (N $_{TM}$, S $_{TM}$) coordinate values. In the Southern Hemisphere, a ten million meter (10,000,000 m) offset must be added to account for the origin. The UTM eastings (E $_{UTM}$) are derived by multiplying the TM eastings (E $_{TM}$) by the scale factor of the central meridian (0.9996) and adding a five-hundred thousand meter (500,000 m) offset to account for the origin. UTM coordinates are always expressed in meters.

UTM Northings, Southings, and Eastings

Northern Hemisphere:

$$N_{UTM} = (0.9996) N_{TM}$$

Southern Hemisphere:

$$S_{UTM} = (0.9996) S_{TM} + 10,000,000 \text{ m}$$

$$E_{UTM} = (0.9996) E_{TM} + 500,000 \text{ m}$$

The UTM zone (Z) can be calculated from the geodetic longitude of the point (converted to decimal degrees). In the continental United States, UTM zones range from ten (10) to nineteen (19).

UTM Zone:

$$Z = (180 + \lambda) / 6 \qquad \text{(east longitude)}$$

$$Z = (180 - \lambda) / 6 \qquad \text{(west longitude)}$$

where

$$Z = \text{UTM zone number}$$

If the computed zone value Z results in a decimal quantity, then the zone must be incremented by one whole zone number.

Example of UTM Zone Calculation:

$$\lambda = 77^d \ 08^m \ 44.3456^s \ W$$

$$Z = 17.14239$$
$$Z = 17 + 1$$
$$Z = 18$$

In the example above, Z is a decimal quantity, therefore, the zone equals seventeen (17) plus one (1).

4-5. Datum Transformations

a. General. Federal Geodetic Control Subcommittee (FGCS) members, which includes USACE have adopted NAD 83 as the standard horizontal datum for surveying and mapping activities performed or financed by the Federal government. To the extent practicable, legally allowable, and feasible, USACE should use NAD 83 in its surveying and mapping activities. Transformations between NAD 27 coordinates and NAD 83 coordinates are generally obtained using the CORPS Convert (i.e., CORPSCON) software package or other North American Datum Conversion (i.e., NADCON) based programs.

b. Conversion techniques. USACE survey control published in the NGS control point database has been already converted to NAD 83 values. However, most USACE survey control was not originally in the NGS database and was not included in the NGS readjustment and redefinition of the national geodetic network. Therefore, USACE will have to convert this control to NAD 83. Coordinate conversion methods considered applicable to USACE projects are discussed below.

(1) Resurvey from NAD 83 Control. A new survey using NGS published NAD 83 control could be performed over the entire project. This could be either a newly authorized project or one undergoing major renovation or maintenance. Resurvey of an existing project must tie into all monumented points. Although this is not a datum transformation technique, and would not normally be economically justified unless major renovation work is being performed, it can be used if existing NAD 27 control is of low density or accuracy.

(2) Readjustment of Survey. If the original project control survey was connected to NGS control stations, the survey may be readjusted using the NAD 83 coordinates instead of the NAD 27 coordinates originally used. This method involves locating the original field notes and observations, and completely readjusting the survey and fixing the published NAD 83 control coordinates.

(3) Mathematical Transformations. Since neither of the above methods can be economically justified on most USACE projects, mathematical approximation techniques for transforming project control data to NAD 83 have been developed. These methods yield results which are normally within ± 1 foot of the actual values and the distribution of errors are typically consistent within a local project area. Since these coordinate transformation techniques involve approximations, they should be used with caution when real property demarcation points and precise surveying projects are involved. When mathematical transformations are employed they should be adequately noted so that users will be aware of the conversion method.

4-6. Horizontal Datum Transformations

a. General. Coordinate transformations from one geodetic reference system to another can be most practically made by using either a local seven-parameter transformation, or by interpolation of datum shift values across a given region.

b. Seven parameter transformations. For worldwide (OCONUS) and local datum transformations, the procedures referenced in USATEC SR-7 1996, "Handbook for Transformation of Datums, Projections, Grids and Common Coordinate Systems" should be consulted. This document contains references for making generalized datum shifts and working with a variety of commonly used map projections.

c. Grid-shift transformations. Current methods for interpolation of datum shift values use the difference between known coordinates of common points from both the NAD 27 and NAD 83 adjustments to model a best-fit shift in the regions surrounding common points. A grid of approximate datum shift values is established based on the computed shift values at common points in the geodetic network. The datum shift values of an unknown point within a given grid square are interpolated along each axis to compute an approximate shift value between NAD 27 and NAD 83. Any point that has been converted by such a transformation method, should be considered as having only approximate NAD 83 coordinates.

d. NADCON/CORPSCON. NGS developed the transformation program NADCON, which yields consistent NAD 27 to NAD 83 coordinate transformation results over a regional area. This technique is based on the above grid-shift interpolation approximation. NADCON was reconfigured into a more comprehensive program called CORPSCON. This software converts between:

NAD 27	NAD 83	SPCS 27	SPCS 83
UTM 27	UTM 83	NGVD 29	NAVD 88
GEOID96	HARN		

Technical documentation and operating instructions for CORPSCON are listed in Appendix B. Since the overall CORPSCON datum shift (from point to point) varies throughout North America, the amount of datum shift across a local project is also not constant. The variation can be as much as 0.1 foot per mile. Some typical NAD 27 to NAD 83 based coordinate shift variations that can be expected over a 10,000 foot section of a project are shown below:

Project Area	SPCS Reference	Per 10,000 feet
Baltimore, MD	1900	0.16 ft
Los Angeles, CA	0405	0.15 ft
Mississippi Gulf Coast	2301	0.08 ft
Mississippi River (IL)	1202	0.12 ft
New Orleans, LA	1702	0.22 ft
Norfolk, VA	4502	0.08 ft
San Francisco, CA	0402	0.12 ft
Savannah, GA	1001	0.12 ft
Seattle, WA	4601	0.10 ft

Such local scale changes will cause project alignment data to distort by unequal amounts. Thus, a 10,000-foot tangent on NAD 27 project coordinates could end up as 9,999.91 feet after mathematical transformation to NAD 83 coordinates. Although such differences may not be appear significant from a lower-order construction survey standpoint, the potential for such errors must be recognized. Therefore, the transformations will not only significantly change absolute coordinates on a project, the datum transformation process will slightly modify the project's design dimensions and/or construction orientation and scale. On a navigation project, for example, an 800.00 foot wide channel could vary from 799.98 to 800.04 feet along its reach, and also affect grid azimuths. Moreover, if the local SPCS 83 grid was further modified, then even larger dimension changes can result. Correcting for distortions may require recomputation of coordinates after conversion to ensure original project dimensions and alignment data remain intact. This is particularly important for property and boundary surveys. A less accurate alternative is to compute a fixed shift to be applied to all data points over a limited area. Determining the maximum area over which such a fixed shift can be applied is important. Computing a fixed conversion factor with CORPSCON can be made to within ±1 foot. Typically, this fixed conversion would be computed at the center of a sheet or at the center of a project and the conversions in X and Y from NAD 27 to NAD 83 and from SPCS 27 to SPCS 83 indicated by notes on the sheets or data sets. Since the conversion is not constant over a given area, the fixed conversion amounts must be explained in the note. The magnitude of the conversion factor change across a sheet is a function of location and the drawing scale. Whether the magnitude of the distortion is significant depends on the nature of the project. For example, a 0.5-foot variation on an offshore navigation project may be acceptable for converting depth sounding locations, whereas a 0.1 foot change may be intolerable for construction layout on an installation. In any event, the magnitude of this gradient should be computed by CORPSCON at each end (or corners) of a sheet or project. If the conversion factor variation exceeds the allowable tolerances, then a fixed conversion factor should not be used. Two examples of using Fixed Conversion Factors follow:

(1) Example 1. Assume we have a 1" = 40' scale site plan map on existing SPCS 27 (VA South Zone 4502). Using CORPSCON, convert existing SPCS 27 coordinates at the sheet center and corners to SPCS 83 (US Survey Foot), and compare SPCS 83-27 differences.

	SPCS 83	SPCS 27	SPCS 83 - SPCS 27
Center of Sheet	N 3,527,095.554 E 11,921,022.711	Y 246,200.000 X 2,438,025.000	dY = 3,280,895.554 dX = 9,482,997.711
NW Corner	N 3,527,595.553 E 11,920,522.693	Y 246,700.000 X 2,437,525.000	dY = 3,280,895.553 dX = 9,482,997.693
NE Corner	N 3,527,595.556 E 11,921,522.691	Y 246,700.000 X 2,438,525.000	dY = 3,280,895.556 dX = 9,482,997.691
SE Corner	N 3,526,595.535 E 11,921,522.702	Y 245,700.000 X 2,438,525.000	dY = 3,280,895.535 dX = 9,482,997.702
SW Corner	N 3,526,595.535 E 11,920,522.704	Y 245,700.000 X 2,437,525.000	dY = 3,280,895.535 dX = 9,482,997.704

Since coordinate differences do not exceed 0.03 feet in either the X or Y direction, the computed SPCS 83-27 coordinate differences at the center of the sheet may be used as a fixed conversion factor to be applied to all existing SPCS 27 coordinates on this drawing.

(2) Example 2. Assuming a 1" = 1,000' base map is prepared of the same general area, a standard drawing will cover some 30,000 feet in an east-west direction. Computing SPCS 83-27 differences along this alignment yields the following:

	SPCS 83	SPCS 27	SPCS 83 - SPCS 27
West End	N 3,527,095.554 E 11,921,022.711	Y 246,200.000 X 2,438,025.000	dY = 3,280,895.554 dX = 9,482,997.711
East End	N 3,527,095.364 E 11,951,022.104	Y 246,200.000 X 2,468,025.000	dY = 3,280,895.364 dX = 9,482,997.104

The conversion factor gradient across this sheet is about 0.2 feet in Y and 0.6 feet in X. Such small changes are not significant at the plot scale of 1" = 1,000'; however, for referencing basic design or construction control, applying a fixed shift across an area of this size is not recommended -- individual points should be transformed separately. If this 30,000-foot distance were a navigation project, then a fixed conversion factor computed at the center of the sheet would suffice for all bathymetric features. Caution should be exercised when converting portions of projects or military installations or projects that are adjacent to other projects that may not be converted. If the same monumented control points are used for several projects or parts of the same project, different datums for the two projects or parts thereof could lead to surveying and mapping errors, misalignment at the junctions and layout problems during construction.

   e. Dual grids ticks. Depicting both NAD 27 and NAD 83 grid ticks and coordinate systems on maps and drawings should be avoided where possible. This is often confusing and can increase the chance for errors during design and construction. However, where use of dual grid ticks and coordinate systems is unavoidable, only secondary grid ticks in the margins will be permitted.

   f. Global Positioning System (GPS). GPS surveying techniques and computations are based on WGS 84 coordinates, which are highly consistent with NAD 83. Differential (static) GPS surveying techniques are accurate for high order control over very large distances. If GPS is used to set new control points referenced to higher order control many miles from the project, inconsistent data may result at the project site. If the new control is near older control points that have been converted to NAD 83, two slightly different network solutions can result, even though both have NAD 83 coordinates. In order to avoid this situation, locate the GPS base stations on the control in the project area, (i.e., don't transfer it in from outside the area). Use the CORPSCON program to convert the old control from NAD 27 to NAD 83 and use these NAD 83 values to initiate the GPS survey. This allows GPS to produce coordinates that are both referenced to NAD 83 and consistent with the old control.

   g. Local project datums. Local project datums that are not referenced to NAD 27 cannot be mathematically converted to NAD 83 with CORPSCON. Field surveys connecting them to other stations that are referenced to NAD 83 are required.

4-7. Horizontal Transition Plan

   a. General. Not all maps, engineering site drawings, documents and associated products containing coordinate information will require conversion to NAD 83. To insure an orderly and timely transition to NAD 83 is achieved for the appropriate products, the following general guidelines should be followed:

   (1) Initial surveys. All initial surveys should be referenced to NAD 83.

   (2) Active projects. Active projects where maps, site drawings or coordinate information are provided to non-USACE users (e.g., NOAA, USCG, FEMA and others in the public and private sector) coordinates should be converted to NAD 83 the next time the project is surveyed or maps or site drawings are updated for other reasons.

   (3) Inactive projects. For inactive projects or active projects where maps, site drawings or coordinate information are not normally provided to non-USACE users, conversion to NAD 83 is optional.

   (4) Datum notes. Whenever maps, site drawings or coordinate information (regardless of type) are provided to non-USACE users, it should contain a datum note, such as the following:

   THE COORDINATES SHOWN ARE REFERENCED TO NAD *[27/83] AND ARE IN
   FEET BASED ON THE SPCS *[27/83] *[STATE, ZONE]. DIFFERENCES
   BETWEEN NAD 27 AND NAD 83 AT THE CENTER OF THE *[SHEET/DATASET]
   ARE *[dLat, dLon, dX, dY]. DATUM CONVERSION WAS PERFORMED USING
   THE COMPUTER PROGRAM "CORPSCON." METRIC CONVERSIONS WERE

BASED ON THE *[US SURVEY FOOT = 1200/3937 METER] [INTERNATIONAL FOOT = 30.48/100 METER].

   b. *Levels of effort.* For maps and site drawings the conversion process entails one of three levels of effort:

   (1) conversion of coordinates of all mapped details to NAD 83, and redrawing the map,

   (2) replace the existing map grid with a NAD 83 grid,

   (3) simply adding a datum note.

For surveyed points, control stations, alignment, and other coordinated information, conversion must be made through either a mathematical transformation or through readjustment of survey observations.

   c. *Detailed instructions.*

   (1) Initial surveys on Civil Works projects. The project control should be established on NAD 83 relative to NGS National Geodetic Reference System (NGRS) using conventional or GPS surveying procedures. The local SPCS 83 grid should be used on all maps and site drawings. All planning and design activities should then be based on the SPCS 83 grid. This includes supplemental site plan mapping, core borings, project design and alignment, construction layout and payment surveys, and applicable boundary or property surveys. All maps and site drawings shall contain datum notes. If the local sponsor requires the use of NAD 27 for continuity with other projects that have not yet converted to NAD 83, conversion to NAD 27 could be performed using the CORPSCON transformation techniques described in Appendix B.

   (2) Active Civil Works Operations and Maintenance projects undergoing maintenance or repair. These projects should be converted to NAD 83 during the next maintenance or repair cycle in the same manner as for newly initiated civil works projects. However, if resources are not available for this level of effort, either redraw the grids or add the necessary datum notes. Plans should be made for the full conversion during a later maintenance or repair cycle when resources can be made available.

   (3) Military Construction and master planning projects. All installations and master planning projects should remain on NAD 27 or the current local datum until a thoroughly coordinated effort can be arranged with the MACOM and installation. An entire installation's control network should be transformed simultaneously to avoid different datums on the same installation. The respective MACOMs are responsible for this decision. However, military operations may require NAD 83, including SPCS 83 or UTM metric grid systems. If so, these shall be performed separate from facility engineering support. A dual grid system may be required for such operational applications when there is overlap with normal facilities engineering functions. Coordinate transformations throughout an installation can be computed using the procedure described in Appendix B. Care must be taken when using transformations from NAD 27 with new control set using GPS methods from points remote from the installation. Installation boundary surveys should adhere to those outlined under real estate surveys listed below.

   (4) Real Estate. Surveys, maps and plats prepared in support of civil works and military real estate activities should conform as much as possible to state requirements. Since most states have adopted NAD 83, most new boundary and property surveys should be based on NAD 83. The local authorities should be contacted before conducting boundary and property surveys to ascertain their policies. It should be noted that several states have adopted the International Foot for their standard conversion from meters to feet. In order to avoid dual coordinates on USACE survey control points that

have multiple uses, all control should be based on the US Survey Foot, including control for boundary and property surveys. In states where the International foot is the only accepted standard for boundary and property surveys, conversion of these points to NAD 83 should be based on the International foot, while the control remains based on the US Survey foot.

(5) Regulatory functions. Surveys, maps and site drawings prepared in support of regulatory functions should begin to be referenced to NAD 83 unless there is some compelling reason to remain on NAD 27 or locally used datum. Conversion of existing surveys, maps and drawing to NAD 83 is not necessary. Existing surveys, maps and drawings need only have the datum note added before distribution to non-USACE users. The requirements of local, state and other Federal permitting agencies should be ascertained before site specific conversions are undertaken. If states require conversions based on the International foot, the same procedures as described above for Real Estate surveys should be followed.

(6) Other existing projects. Other existing projects, e.g., beach nourishment, submerged offshore disposal areas, historical preservation projects, etc., need not be converted to NAD 83. However, existing surveys, maps and drawings should have the datum note added before distribution to non-USACE users.

(7) Work for others. Existing projects for other agencies will remain on NAD 27 or the current local datum until a thoroughly coordinated effort can be arranged with the sponsoring agency. The decision to convert rests with the sponsoring agency. However, existing surveys, maps and drawings should have the datum note added before distribution to non-USACE users. If sponsoring agencies do not indicate a preference for new projects, NAD 83 should be used. The same procedures as described above for Initial Surveys on Civil Works Projects should be followed.

4-8. Vertical Datums

a. General. A vertical datum is the surface to which elevations or depths are referred to or referenced. There are many vertical datums used within CONUS. The surveyor should be aware of the vertical control datum being used and its practicability to meet project requirements. Further technical details on vertical datums are in Appendix C, Development and Implementation of NAVD 88.

b. NGVD 29. NGVD 29 was established by the United States Coast and Geodetic Survey (USC&GS) 1929 General Adjustment by constraining the combined US and Canadian First Order leveling nets to conform to Mean Sea Level (MSL). It was determined at 26 long term tidal gage stations that were spaced along the east and west coast of North American and along the Gulf of Mexico, with 21 stations in the US and 5 stations in Canada. NGVD 29 was originally named the Mean Sea Level Datum of 1929. It was known at the time that the MSL determinations at the tide gages would not define a single equipotential surface because of the variation of ocean currents, prevailing winds, barometric pressures and other physical causes. The name of the datum was changed from the Mean Sea Level Datum to the NGVD 29 in 1973 to eliminate the reference to sea level in the title. This was a change in name only; the definition of the datum established in 1929 was not changed. Since NGVD 29 was established, it has become obvious that the geoid based upon local mean tidal observations would change with each measurement cycle. Estimating the geoid based upon the constantly changing tides does not provide a stable estimate of the shape of the geoid.

c. NAVD 88. NAVD 88 is the international vertical datum adopted for use in Canada, the United States and Mexico. NAVD 88 is based on gravity measurements made at observation points in the network and only one datum point, at Pointe-au-Pere/Rimouski, Quebec, Canada, is used. The vertical reference surface is therefore defined by the surface on which the gravity values are equal to the control point value. The result of this adjustment is newly published NAVD 88 elevation values for benchmarks (BM) in the NGS inventory. Most Third Order benchmarks, including those of other Federal, state and local government agencies, were not included in the NAVD 88 adjustment. The Federal Geodetic Control Subcommittee (FGCS) of the Federal Geographic Data Committee (FGDC) has affirmed that NAVD 88 shall be the official vertical reference datum for the US. The FGDC has prescribed that all surveying and mapping activities performed or financed by the Federal Government make every effort to begin an orderly transition to NAVD 88, where practicable and feasible. Procedures for performing this transition are outlined in Appendix E.

d. Mean Sea Level datums. Some vertical datums are referenced to mean seal level. Such datums typically are maintained locally or within a specific project area. The theoretical basis for these datums is local mean sea level. Local MSL is a vertical datum based on observations from one or more tidal gaging stations. NGVD 29 was based upon the assumption that local MSL at 21 tidal stations in the US and 5 tidal stations in Canada equaled 0.0000 foot on NGVD 29. The value of MSL as measured over the Metonic cycle of 19 years shows that this assumption is not valid and that MSL varies from station to station.

e. Lake and tidal datums. Some vertical datums are referenced to tidal waters or lake levels. An example of a lake level used as a vertical datum is the International Great Lakes Datum of 1955 (IGLD 55), maintained and used for vertical control in the Great Lakes region of CONUS. These datums undergo periodic adjustment. For example, the IGLD 55 was adjusted in 1985 to produce IGLD 85. IGLD 85 has been directly referenced to NAVD 88 and originates at the same point as NAVD 88. Tidal datums typically are defined by the phase of the tide and usually are described as mean high water, mean low water, and mean lower low water. For further information on these and other tidal datum related terms, the reader is advised to refer to Appendix D (Requirements and Procedures for Referencing Coastal Navigation Projects to Mean Lower Low Water (MLLW) Datum) and EM 1110-2-1003 (Hydrographic Surveying).

f. Other vertical datums. Other areas may maintain and employ specialized vertical datums. For instance, vertical datums maintained in Alaska, Puerto Rico, Hawaii, the Virgin Islands, Guam, and other islands and project areas. Specifications and other information for these particular vertical datums can be obtained from the particular FOA responsible for survey related activities in these areas, or the National Ocean Service (NOS).

4-9. Vertical Datum Transformations

a. General. There are several reasons for USACE commands to convert to NAVD 88.

(1) Differential leveling surveys will close better.

(2) NAVD 88 will provide a reference to estimate GPS derived orthometric heights.

(3) NAVD 88 Height values will be available in convenient form from the NGS database.

(4) Federal surveying and mapping agencies will stop publishing on NGVD 29.

(5) NAVD 88 is recommended by ACSM and FGCS.

(6) Surveys performed for the Federal government will require use of NAVD 88.

The conversion process entails one of two levels of effort:

(1) conversion of all elevations to NAVD 88 and redrawing the map,

(2) adding a datum note based on an approximate conversion.

b. VERTCON. VERTCON is a software program developed by NGS that converts elevation data from NGVD 29 to NAVD 88. Although the VERTCON software has been fully incorporated into the software application package CORPSCON, it will be referred to below as a separate program. VERTCON uses benchmark heights to model the shift between NGVD 29 and NAVD 88 that is applicable to a given area. In general, it is only sufficiently accurate to meet small-scale mapping requirements. VERTCON should not be used for converting benchmark elevations used for site plan design or construction applications. Users input the latitude and longitude for a point and the vertical datum shift between NGVD 29 and NAVD 88 is reported. The root-mean-square (RMS) error of NGVD 29 to NAVD 88 conversion, when compared to the stations used to create the conversion model, is ±1 cm; with an estimated maximum error of ± 2.5 cm. Depending on network design and terrain relief, larger differences (e.g., 5 to 50 cm) may occur the further a bench mark is located from the control points used to establish the model coefficients. For this reason, VERTCON should only be used for approximate conversions where these potential errors are not critical.

c. Datum note. Whenever maps, site drawings or spatial elevation data are provided to non-USACE users, they should contain a datum note that provides, at minimum, the following information:

The elevations shown are referenced to the *[NGVD 29] [NAVD 88] and are in *[feet] [meters]. Differences between NGVD 29 and NAVD 88 at the center of the project sheet/data set are shown on the diagram below. Datum conversion was performed using the *[program VERTCON] [direct leveling connections with published NGS benchmarks] [other]. Metric conversions are based on *[US Survey Foot = 1200/3937 meters] [International Survey Foot = 0.3048 meters].

4-10. Vertical Transition Plan

a. General. A change in the accepted vertical datum will affect USACE engineering, construction, planning, and surveying activities. The cost of conversion could be substantial at the onset. There is a potential for errors in conversions inadvertently occurring. The effects of the vertical datum change can be minimized if the change is gradually applied over time; being applied to future projects and efforts, rather than concentrated on changing already published products. In order to insure an orderly and timely transition to NAVD 88 is achieved for the appropriate products, the following general guidelines should be followed.

b. Conversion criteria. Maps, engineering site drawings, documents and associated spatial data products containing elevation data may require conversion to NAVD 88. Specific requirements for conversion will, in large part, be based on local usage -- that of the local sponsor, installation, etc. Where applicable and appropriate, this conversion should be recommended to local interests.

c. Newly authorized construction projects. Generally, initial surveys of newly authorized projects should be referenced to NAVD 88. In addition to design/construction, this would include wide-area master plan mapping work. The project control should be referenced to NAVD 88 using conventional or GPS surveying techniques. All planning and design activities should be based upon NAVD 88. All maps and site drawings shall contain datum notes as described below. If the sponsor/installation requires the use of NGVD 29 or some other local vertical reference datum for continuity, the relationship between NGVD 29 and NAVD 88 shall be clearly noted on all maps, engineering site drawings, documents and associated products.

d. Active projects. On active projects where maps, site drawings, or elevation data are provided to non-USACE users, the conversion to NAVD 88 should be performed. This conversion to NAVD 88 may be performed the next time the project is surveyed or when the maps/site drawings are updated for other reasons. Civil works projects may be converted to NAVD 88 during the next maintenance or repair cycle in the same manner as for newly initiated civil works projects. However, if resources are not available for this level of effort, redraw the maps or drawings and add the necessary datum note. Plans should be made for the full conversion during a later maintenance or repair cycle when resources can be made available. Military installations should remain on NGVD 29 or the local vertical datum until a thoroughly coordinated effort can be arranged with the MACOM and installation. An entire installation's control network should be transformed simultaneously to avoid different datums on the same installation. MACOMs should be encouraged to convert to NAVD 88. However, the respective MACOMs are responsible for this decision.

e. Inactive projects. For inactive projects or active projects where maps, site drawings or elevation data are not normally provided to non-USACE users, conversion to NAVD 88 is optional.

f. Work for others. Projects for other agencies will remain on NGVD 29 or the current local vertical datum until a thoroughly coordinated effort can be arranged with the sponsoring agency. Other agencies should be encouraged to convert their projects to NAVD 88, although the decision to convert rests with the sponsoring agency. However, surveys, maps and drawings should have the datum note described below added before distribution to non-USACE users. If sponsoring agencies do not indicate a preference for new projects, NAVD 88 should be used.

g. Miscellaneous projects. Other projects referenced to strictly local datum, such as, beach nourishment, submerged offshore disposal areas, historical preservation projects, etc., need not necessarily be converted to NAVD 88. However, it is recommended that surveys, maps and drawings have a clear datum reference note added before distribution to non-USACE users.

h. Real Estate. Surveys, maps, and plats prepared in support of civil works and military real estate activities should conform as much as possible to state requirements. Many states are expected to adopt NAVD 88 (by statute) as an official vertical reference datum. This likewise will entail a transition to NAVD 88 in those states. State and local authorities should therefore be contacted to ascertain their current policies. Note that several states have adopted the International Foot for their standard conversion from meters to feet. In order to avoid dual elevations on USACE survey control points that have multiple uses, it is recommended that published elevations be based on the US Survey Foot. In states where the International Foot is the only accepted standard for boundary and property surveys, conversion of these elevations to NAVD 88 should be based on the International Foot while the control remains based on the US Survey Foot.

4-11. Mandatory Requirements

Horizontal and vertical datum transition criteria in paragraphs 4-7 and 4-10 are mandatory.

Chapter 5
Horizontal Control Survey Techniques

5-1. Introduction

a. General. Primary horizontal control (Third Order Class I or higher) is established to serve as a basic framework for large mapping projects, to establish new horizontal control in a remote area, or to further densify existing horizontal control in an area.

b. Instruments. Minimum instrument requirements for the establishment of primary control will typically include a repeating theodolite having an optical micrometer with a least-count resolution of six seconds (i.e., 6") or better; a directional theodolite having an optical micrometer with a least count resolution of one arc-second; an EDM capable of a resolution of 1:10,000; or a total station having capabilities comparable to, or better than, any of the instruments just detailed.

c. Monumentation. Primary horizontal control points not permanently monumented in accordance with criteria and guidance established in EM 1110-1-1002 should meet the following minimum standards:

(1) Markers. Primary horizontal control points shall be marked with semi-permanent type markers (e.g., re-bar, railroad spikes, or large spikes). If concrete monuments are required, they will be set prior to horizontal survey work. These monuments will be established in accordance with EM 1110-1-1002.

(2) Installation. Primary horizontal control points shall be placed either flush with the existing ground level or buried a minimum of one-tenth of a foot below the surface.

(3) Reference marks. Each primary control point should be referenced by a minimum of two points to aid in future recovery of that point. For this reference, well defined natural or manmade objects may be used. The reference point(s) can be either set or existing and should be within one hundred feet of the control point.

(4) Sketches. A sketch shall be placed in a standard field survey book. The sketch, at minimum, will show the relative location of each control point to the reference points and major physical features within one hundred feet of the point.

d. Redundancy. A minimum of three to four repeated angle measurements will be made for establishing primary control points--one angle and one distance will not be sufficient. When using high precision total stations, only half as many readings are generally required (two data set collections). With EDM distance measurements, a minimum of two readings shall be taken at each setup and recorded in a standard field book. The leveled height of the instrument and the height of the reflector shall be measured carefully to within 0.01 foot and recorded in the field book. Each slope distance shall be reduced to a horizontal distance using either reciprocal vertical angle observations or from the known elevation of each point obtained from differential leveling.

e. Repeating theodolite. If a repeating theodolite (e.g., Wild T1) is used for the horizontal angles, the instrument will be pointed at the backsight station with the telescope in a direct reading position, and the horizontal vernier set to zero degrees. All angles shall then be turned to the right, and the first angle recorded in a field book. The angle shall be repeated a minimum of four times (i.e., two sets) by alternating the telescope and pointing in the direct and inverted positions. The last angle will also be recorded in the field book. If the first angle deviates more than five seconds (5") from the result of the last angle divided by four, the process shall be repeated until the deviation is less than or equal to five seconds. Multiples of 360 degrees may need to be added to the last angle before averaging. The horizon shall be closed by repeating this process for all of the sights to be observed from that location. The foresight for the last observation shall be the same as the backsight for the first observation. If the sum of all the angles turned at any station deviates more than ten seconds (10") from 360 degrees, the angles shall be turned again until the summation is within this tolerance.

f. Directional theodolite. If a directional theodolite (e.g., Wild T2, Wild T3) is used for the horizontal angles, the instrument shall be pointed at the backsight station with the telescope in a direct reading position and the horizontal scales set to within ten seconds (10") of zero degrees. The scales shall be brought into coincidence and the angle read and recorded in the field book. The angles shall then be turned to each foresight in a clockwise direction, and the angles read and recorded in a field book. This process will continue in a clockwise direction and shall include all sights to be observed from that station. The telescope shall then be inverted and the process repeated in reverse order, except the scales are not to be reset, but will be read where it was originally set. The angles between stations may then be computed by differencing the direct and reverse readings. This process shall be repeated three times for a total of three direction set readings.

g. Horizontal distances. To reduce EDM slope distances to horizontal, a vertical angle observation must be obtained from each end of each line being measured. The vertical angles shall be read in both the direct and inverted scope positions and adjusted. If the elevations for the point on each end of the line being measured are obtained by differential leveling, then this vertical angle requirement is not necessary.

h. Targets. All targets established for backsights and foresights shall be centered directly over the measured point. Target sights may be a reflector or other type of target set in a tribrach, a line rod plumbed over the point in a tripod, or guyed in place from at least three positions. Artificial sights (e.g., a tree on the hill behind the point) or hand held sights (e.g., line rod or plumb bob string) will not be used to set primary control targets.

e. Calibration. All total stations, EDM, and prisms used for primary control work shall be serviced regularly and checked frequently over lines of known length. Calibration should be done at least annually. Theodolite instruments should be adjusted for collimation error at least once a year and whenever the difference between direct and reverse reading of any theodolite deviates more than thirty seconds from 180 degrees. Readjustment of the cross hairs and the level bubble should be done whenever misadjustments affect the instrument reading by more than the least count of the reading scales of the theodolite.

5-2. Secondary Horizontal Control

a. General. Secondary horizontal control (Third Order Class II or lower) is established to determine the location of structure sections, cross sections, or topographic surfaces, or to pre-mark requirements for small to medium scale photogrammetric mapping.

b. Requirements. Secondary horizontal control requirements are identical to that described for primary horizontal control with the following exceptions.

(1) Monumentation. It is not required for secondary horizontal control points to have two reference points.

(2) Occupation. Secondary horizontal control points can be established by one angle and one distance.

(3) When a total station or EDM is used, a minimum of two readings shall be taken at each setup and recorded in a standard field book.

(4) If a repeating theodolite is used for the horizontal angles, the angle measurement shall be repeated a minimum of two times by alternating the telescope and pointing in the direct and inverted positions.

(5) If a directional theodolite is used for the horizontal angles, the process (described for primary control) shall be repeated two times for a total of two data set collections.

5-3. Traverse Survey Standards

a. General. A survey traverse is defined as the measurement of the lengths and directions of a series of straight lines connecting a series of points on the earth. Points connected by the lines of traverse are known as traverse stations. The measurements of the lengths and directions are used to compute the relative horizontal positions of these stations. Traversing is used for establishing basic area control surveys where observation of horizontal directions and distances between traverse stations, and elevations of the stations, must be determined. Astronomic observations and GPS surveys are made along a traverse at prescribed intervals to control the azimuth of the traverse. The interval and type of astronomic observation will depend upon the order of accuracy required and the traverse methods used.

b. Traverse types. There are two basic types of traverses, namely, closed traverses and open traverses.

(1) Closed traverse. A traverse that starts and terminates at a station of known position is called a closed traverse. The order of accuracy of a closed traverse depends upon the accuracy of the starting and ending known positions and the survey methods used for the field measurements. There are two types of closed traverses.

(a) Loop traverse. A loop traverse starts on a station of known position and terminates on the same station. An examination of the position misclosure in a loop traverse will reveal measurement blunders and internal loop errors, but will not disclose systematic errors or external inaccuracies in the control point coordinates.

(b) Connecting traverse. A connecting traverse starts on a station of known position and terminates on a different station of known position. When using this type of traverse the systematic errors and position inaccuracies can be detected and eliminated along with blunders and accidental errors. The ability to correct measurement error depends on the known accuracy of the control point coordinates.

(2) Open traverse. An open traverse starts on a station of known position and terminates on a station of unknown position. With an open traverse, there are no checks to determine blunders, accidental errors, or systematic errors that may occur in the measurements. The open traverse is very seldom used in topographic surveying because a loop traverse can usually be accomplished with little added expense or effort.

c. Requirements. The following minimum guidelines should be followed for traverse procedures:

(1) Origin. All traverses will originate from and tie into an existing control line of equal or higher accuracy.

(a) Astronomic observation. If it is impossible to start or terminate on stations of known position and/or azimuth, then an astronomic observation for position and/or azimuth must be conducted. For Third Order surveys, astronomic azimuth observations are made at intervals along the traverse and at abrupt changes in the direction of the traverse. The placement of these astronomic stations is governed by the order of accuracy required.

(b) Traverse setup. The specific route of a new traverse shall be selected with care, keeping in mind its primary purpose and the flexibility of its future use. Angle points should be set in protected locations if possible. Examples of protected locations include fence lines, under communication or power lines, near poles, or near any permanent concrete structure. It may be necessary to set critical points below the ground surface. If this is the case, reference the traverse point relative to permanent features by a sketch, as buried points are often difficult to recover at future dates.

(c). Accuracy. Traversing is conducted under four general orders of accuracy:

- First Order
- Second Order
- Third Order
- Fourth Order

The order of accuracy for traversing is determined by the equipment and methods used to collect the traverse measurements, by the final accuracy attained, and by the coordinate accuracy of the starting and terminating stations of the traverse. The point closure standards in either Table 3-1 or FGCS 1984 must be met for the appropriate accuracy classification to be achieved.

5-4. Traverse Survey Guidelines

a. General. Survey traverse work involves several basic steps to plan and execute.

- researching existing control in the project area
- design survey to meet specifications
- determine types of measurements
- determine types of instruments
- determine field procedures
- site reconnaissance and approximate surveys
- install monuments and traverse stations
- data collection
- data reduction
- data adjustment
- prepare survey report

`b. Control traverses.* Control traverses are run for use in connection with all future surveys to be made in the area of consideration. They may be of First, Second, Third, or Fourth Order accuracy, depending on project requirements.

(1) Preparation. Most project requirements will be satisfied with Second or Third Order accuracies. For a Second Order traverse, it is recommended that permanent points be established at

intervals of one mile or less, starting at a known point--preferably a National Geodetic Survey (NGS) published control point. Plan the traverse to follow a route that will be centered as much on the project area as possible, and avoiding areas that will be affected by construction, traffic, or other forms of congestion. The route should provide a check into other known points as often as practicable. After determining the route, it is best to then set permanent monuments (e.g., stakes, iron rods, brass caps in concrete, or some other suitable monument) at each angle point and any intermediate points desired. Refer to EM 1110-1-1002 for further guidance on survey markers and monumentation. Ensure there is a clear line of sight from angle point to angle point and determine an organized numbering or naming system to mark all points when set.

(3) Measurements. Manufacturer instructions for operation of the EDM or total station should be followed. When using an EDM or total station, a minimum of two readings will be made before moving to the next occupation point. All readings should agree within the resolution of the instrument or 0.001 foot of the original reading. Determination of angles should be made immediately after distance determination. Special care should be taken with the type of sights used for angle measurement--fixed rigid sights should be used, not hand held targets. For directional theodolite or total station angle measurements, at least three sets of angles should be made. Adequate results can be obtained with fewer angles if precision equipment is used. A horizon closure may be performed as a check.

(3) Reductions. All survey field notes should be carefully and completely reduced with the mean angle calculated in the field and recorded along with the sketch. All traverse adjustments should be made in the office. A sketch of the monument location should be made in the field and a detailed description on how to recover it should be recorded in writing. This information can be used for making subsequent record of the survey monument and survey report.

c. Right of Way traverse. A right of way traverse typically is a Third Order traverse, starting and ending on known points. This type of traverse is usually run with a transit and steel tape, EDM, or total station. The style of notes is similar to most traverses with the only difference being the type of detail shown. Fences are of particular importance in determining right of way limits, especially when working in an area not monumented. Notes for right of way traverses should be especially clear and complete for many times this type of traverse is the basis for legal or court hearings regarding true property corners. If a search for a corner is made and nothing is found, a statement should be written in the field book to this effect. Property title searches and deed research will generally be required to obtain appropriate existing descriptions, plans, and other documents which are generally available in the public record.

d. Stadia traverse. Uses of stadia traverses include rough or reconnaissance type surveys, checking on another traverse for errors, and control for a map being made by stadia methods on a very large scale. A stadia traverse typically is run along a route that will best suit the point location requirements of the survey. The stadia points or stadia angle points will be set at locations that will best recover desired information and will be set in protected locations for future use.

e. Compass traverse. A compass traverse is made to establish the direction of a line by compass measurements (i.e., no angles are turned). Distances are usually measured by stadia or paced.

f. Azimuth traverse. Compass bearings break the circle directions into four quadrants, while an azimuth measures direction from true North, South, or on some other base. Azimuths should always be determined as a right deflection from the base point to the reference object. Astronomical observations, GPS surveys, and specialized instruments such as gyro theodolites and gyro theodolite attachments are used to measure precise azimuths.

5-5. Traverse Classifications

a. *General.* Table 5-1 lists specific traverse requirements necessary to meet Second and Third Order type accuracies.

b. *Second Order traverse.* Second Order traverse is used extensively for subdividing an area between First and Second Order triangulation and First Order traverse. Second Order traverse must originate and terminate on existing First or Second Order control that has been previously adjusted.

c. *Third Order traverse.* Third Order traverse is normally used for detailed topographic mapping. Third Order traverse must start and close on existing control stations of Third or higher order accuracy.

Table 5-1. Traverse Requirements

Requirement	Second Order	Third Order
Horizontal Angles		
Instrument	0.2" - 1.0"	1.0"
Repetitions	6 - 8	2 - 4
Rejection Limit	4" - 5"	5"
Number of Courses between Azimuth Checks		
Steel Tape	25	35 - 50
EDM	12 - 16	25
Azimuth Closure		
Standard error	2.0"	5.0"
Azimuth Closure at Checkpoint		
Azimuth Checkpoint	3" per station	5" per station
or		
Azimuth Checkpoint	$(10")*N^{1/2}$	$(15")*N^{1/2}$
	where N is the number of stations carrying azimuth	

c. *Lower Order.* Traverses of lower than Third Order are used for controlling points when a relatively large error in position is permissible. For example, map compilation requirements for a horizontal control panel point for 1:50,000 mapping shall be located to within 6 meters of its true relationship to the basic control. For 1:25,000 mapping the requirement is usually to within 3 meters. The allowable errors permit accuracies to vary from generally 1 part in 500 to 1 part in 5000, depending on the distance the lower order traverse must travel, the type of control at the start of the traverse, the desired accuracy of the control point, and the methods and equipment used in the traverse. Using Third Order methods should be carefully considered, even though the points are not to be monumented permanently.

5-6. Triangulation and Trilateration

a. *General.* A triangulation network consists of a series of angle measurements that form joined or overlapping triangles in which an occasional baseline distance is measured. The sides of the network are calculated from angles measured at the vertices of the triangle. A trilateration network consists of a series of distance measurements that form joined or overlapped triangles where all the sides of the triangles and only enough angles and directions to establish azimuth are determined.

b. Networks. When practicable, all triangulation and trilateration networks will originate from and tie into existing coordinate control of equal or higher accuracy than the work to be performed. An exception to this would be when performing triangulation or trilateration across a river or some obstacle as part of a chained traverse. In this case, a local baseline should be set. Triangulation and trilateration surveys should have adequate redundancy and are usually adjusted using least squares methods.

c. Accuracy. Point closure standards listed in Table 3-1 must be met for the appropriate accuracy classification to be achieved. If project requirements are higher order, refer also to the FGCS Standards and Specifications for Geodetic Control Networks (FGCS 1984).

d. Resection. Three point resection is a form of triangulation. Three point resection may be used in areas where existing control points cannot be occupied or when the work does not warrant the time and cost of occupying each station. Triangulation of this type will be considered Fourth Order, although Third Order accuracy can be obtained if a strong triangular figure is used and the angles are accurately measured. The following minimum guidelines should be followed when performing a three point resection:

(1) Location. Points for observation should be selected so as to give strong geometric figures such as with angles between 60 and 120 degrees of arc.

(2) Redundancy. If it is possible to sight more than three control points, the extra points should be included in the figure. If possible, occupy one of the control stations as a check on the computations and to increase the positioning accuracy. Occupation of a control station is especially important if it serves as a control of the bearing or direction of a line for a traverse that originates from this same point.

(3) Measurements. Both the interior and exterior angles shall be observed and recorded. The sum of these angles shall not vary by more than 3 arc-seconds per angle from 360 degrees. Each angle will be turned not less than 2-4 times (in direct and inverted positions).

5-7. Bearing and Azimuth Determination

a. Bearing types. The bearing of a line is the direction of the line with respect to a given meridian. A bearing is indicated by the quadrant in which the line falls and the acute angle that the line makes with the meridian in that quadrant. Observed bearings are those for which the actual bearing angles are measured, while calculated bearings are those for which the bearing angles are indirectly obtained by calculations. A true bearing is made with respect to the astronomic north reference meridian. A magnetic bearing is one whose reference meridian is the direction to the magnetic poles. The location of the magnetic poles is constantly changing; therefore the magnetic bearing between two points is not constant over time. The angle between a true meridian and a magnetic meridian at the same point is called its magnetic declination. An assumed bearing is a bearing whose prime meridian is assumed. The relationship between an assumed bearing and the true meridian should be defined, as is the case with most state plane grid coordinate systems.

b. Bearing determination guidelines. All bearings used for engineering applications will be described by degrees, minutes, and seconds in the direction in which the line is progressing. Bearings are recorded with respect to its primary direction, north or south, and next the angle east or west. For example, a line can be described as heading north and deflected so many degrees east or west. Alternatively, a line also can be described as heading south and deflected so many degrees east or west. A bearing will never be listed with a value over 90 deg (i.e., the bearing value always will be between over 0 deg and 90 deg.

c. Azimuth types. The azimuth of a line is its direction as given by the angle between the meridian and the line, measured in a clockwise direction. Azimuths can be referenced from either the south point or the north point of a meridian. Assumed azimuths are often used for making maps and performing traverses, and are determined in a clockwise direction from an assumed meridian. Assumed azimuths are sometimes referred to as "localized grid azimuths". Azimuths can be either observed or calculated. Calculated azimuths consist of adding to or subtracting field observed angles from a known bearing or azimuth to determine a new bearing or azimuth.

d. Azimuth determination guidelines. Azimuths will be determined as a line with a clockwise angle from the north or south end of a true or assumed meridian. For traverse work using angle points, the closure requirements in Table 5-1 will be followed.

e. Astronomic azimuth. In order to control the direction of a traverse, an astronomic azimuth must be observed at specified intervals and abrupt changes of direction of the traverse. Astronomic azimuth observations can be made by the well-known hour angle or altitude methods. Azimuth observations should be divided evenly between the backsight and foresight stations as reference objects. Using the rear station, turn clockwise to forward station then to star, reverse telescope on star, then forward station and back to rear station. Then using forward station, turn clockwise to rear station then to star, reverse telescope on star, then rear station and back to forward station. The number of position repetitions will depend upon the order of accuracy required.

f. Position. For Second Order traverse, the observation of position for a Laplace azimuth will depend upon the use of the traverse. The project instructions typically will specify when an astronomic position is required. For some traverses, it may be necessary to observe astronomic positions to obtain the starting and terminating azimuth data; however, such practices are now largely obsolete goven GPS positioning capabilities.

5-8. Mandatory Requirements

The traverse closure requirements in Table 5-1 are mandatory.

Chapter 6
Vertical Control Survey Techniques

6-1. General

Vertical control is established to provide a basic framework for large mapping projects, to establish new vertical control in remote areas, or to further densify existing vertical control in an area. The purpose of vertical control surveys is to establish elevations at convenient points over the project area. These established points (benchmarks) can then serve as points of departure and closure for leveling operations and as reference benchmarks during subsequent construction work. The NGS, NOS, USGS, other Federal agencies, and many USACE commands have established vertical control throughout the CONUS. Unless otherwise directed, these benchmarks will be used as a basis for all vertical control surveys. Descriptions of benchmark data and their published elevation values can be found in data holdings issued by the agency maintaining the circuit. Information on USACE maintained points can be found at District or Division offices.

a. Differential leveling. With differential leveling, differences in elevation are measured with respect to a horizontal line of sight established by the leveling instrument. Once the instrument is leveled (using either a spirit bubble or automated compensator), its line of sight lies a horizontal plane. Leveling comprises a determination of the difference in height between a known elevation and the instrument and the difference in height from the instrument to an unknown point by measuring the vertical distance with a precise or semi-precise level and leveling rods. Digital (or Bar Code) levels are used to automatically measure, store, and compute heights, and are capable of achieving Second Order or higher accuracies. Manufacturer's procedures should be followed to achieve the point closure standards shown in Table 3-2. When leveling in remote areas where the density of basic vertical control is scarce, the semi-precise rod is generally used. The semi-precise rod should be graduated on the face to centimeters and the back to half-foot intervals. When leveling in urban areas or areas with a high density of vertical control where ties to higher order control are readily available, the standard leveling rods are used--e.g., a Philadelphia rod graduated to hundredths of a foot. Other rods that are graduated to centimeters can be used. Both types of rods are furnished with targets and verniers that will permit reading of the scale to millimeters or thousandths of a foot if required by specifications. This is generally not required on lower order trigonometric level lines. Standard stadia rods may also be used for lower order level lines. The stadia rod is graduated to the nearest 0.05 foot, or two centimeters. These rods are generally equipped with targets or verniers, but if project specifications require, they can be estimated to hundredths of a foot.

b. Trigonometric heighting. This method applies the fundamentals of trigonometry to determine the differences in elevation between two points by observing a horizontal distance and the vertical angles above or below a horizontal plane. Trigonometric leveling is generally used for Second Order or lower order accuracy vertical positioning. Trigonometric leveling is especially effective in establishing control for profile lines, for strip photography, and in areas where the landscape is steep. With trigonometric leveling operations, it is necessary to measure the height of instrument (and target) above the monument, the slope distance, and the vertical angle and the rod intercept. From this data, the vertical difference in elevation can be computed using the sine of the vertical angle and applying the rod difference (Figure 6-1). Refinements to this technique include doubling vertical angles, taking differences from both stations and using the mean values. If the horizontal distance is known between the instrument and the rod, it is not necessary to determine the slope distance. The instrument most commonly used for trigonometric heighting is a directional theodolite or Total Station. Manufacturer specifications and procedures should be followed to achieve the point closure standards in Chapter 3.

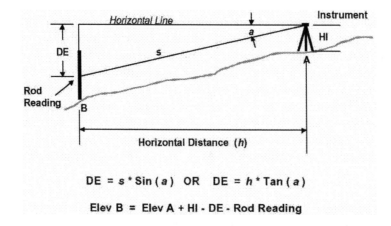

DE = s * Sin (a)   OR   DE = h * Tan (a)

Elev B = Elev A + HI - DE - Rod Reading

Figure 6-1. Trigonometric heighting

   c. Barometric heighting. This method uses the differences in atmospheric pressure as observed with a barometer or altimeter to determine the differences in elevation between points. This method is the least accurate of determining elevations. Because of the lower achievable accuracies, this method should only be used when other methods are not feasible or would involve great expense. Generally, this method is used for elevations when the map scale is to be 1:250,000 or smaller.

   d. Reciprocal leveling. Reciprocal leveling is a method of carrying a level circuit across an area over which it is impossible to run regular differential levels with balanced sights (Figure 6-2). Most level operations require a line of sight less than 300 or 400 feet long. However, it may be necessary to shoot 500-1,000 feet, or even further, in order to span across a river, canyon, or other obstacle. Where such spans must be traversed, reciprocal leveling is appropriate. The reciprocal leveling procedure can be described as follows. Assume points "A" and "B" are turns on opposite sides of the obstacle to be spanned (Figure 6-2) where points A and B are intervisible. Two calibrated rods are used, one at point A, and the other at point B. With the instrument near A, read rod at A, then turn to B and have target set as close as possible and determine the difference in elevation. Leaving rods at A and B, move the instrument around to point B, read B, then turn to read A and again determine the difference in elevation. The mean of the two results is the final height difference to be applied to the elevation of A to get an elevation value for point B. If the long sight is difficult to determine, it is suggested that a target be used and the observations repeated several times to determine an average value. For more precise results it will be necessary to take several foresights, depending on the length of the sight. It is typical to take as many as 20 to 30 sightings. When taking this many sightings, it is critical to relevel the instrument and reset the target after each observation. Reciprocal leveling assumes the conditions during the survey do not change significantly for the two positions of the level. Reciprocal leveling with two instruments should never be done unless both instruments are used on both sides of the obstacle and the mean result of both sets used. The use of two instruments is advised if it is a long trip around the obstacle. Reciprocal leveling is effective only if the instruments used will yield measurements of similar precision.

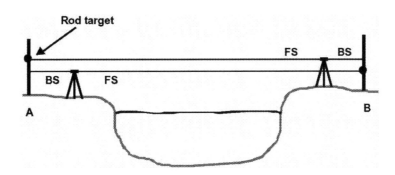

Figure 6-2. Reciprocal leveling for river crossing

    e. Three wire leveling. This method can be used for most types of leveling work and will achieve any practical level of accuracy. However, most applications do not require the accuracies possible with three wire leveling; plus, it is labor intensive. Three wire leveling can be applied if the reticule of the level has stadia lines and substadia that are spaced so that the stadia intercept is 0.3 foot at 100 feet, rather than the more typical 1.0 feet at 100 feet. The substadia lines in instruments meant for three wire leveling are short cross lines that cannot be mistaken for the long central line used for ordinary leveling. Although there are many different observing techniques for three wire leveling, in the following example, the rod is read at each of the three lines and the average is used for the final result. Before each reading, the level bubble is centered. The half-stadia intervals are compared to check for blunders. The following values were taken and recorded and calculations made:

Upper Wire:	8.698	2.155 :Upper Interval
Middle Wire:	6.543	
Lower Wire:	4.392	2.151 :Lower Interval
Sum	19.633	
Average	6.544	

The final rod reading would be 6.544 feet. The upper and lower intercepts differ by only 0.004 foot--an acceptable error for this sort of leveling and evidence that no blunder has been made. It is recommended that "Yard Rods" specifically designed for three wire leveling operations be used instead of Philadelphia rods that are designed for ordinary leveling.

    f. Two rod leveling. In order to increase the productivity in precise leveling operations, it is advisable to use two rods. When the observations are completed at any instrument setup, the rods and the instruments are moved forward simultaneously. An even number of setups should be used to minimize the possible effects of rod index error. Two rods are recommended when using an automatic level, as this takes full advantage of the productivity possible with this type of instrument.

g. Tidal benchmarks and datums. For guidance on the establishment of tidal benchmarks and datums refer to Appendix D in this manual or EM 1110-2-1003.

6-2. Second Order Leveling

a. General. The leveling operation consists of holding a rod vertically on a point of known elevation. A level reading is then made through the telescope on the rod, known as a backsight (BS), which gives the vertical distance from the ground elevation to the line of sight. By adding this backsight reading to the known elevation, the line of sight elevation, called "height of instrument" (HI), is determined. Another rod is place on a point of unknown elevation, and a foresight (FS) reading is taken. By subtracting the FS reading from the height of instrument, the elevation of the new point is established. After the foresight is completed, the rod remains on that point and the instrument and back rod are moved to forward positions. The instrument is set up approximately midway between the old and new rod positions. The new sighting on the back rod is a backsight for a new HI, and the sighting on the front rod is a FS for a new elevation. The points on which the rods are held for foresights and backsights are called "turning points." Other foresights made to points not along the main line are known as "sideshots." This procedure is used as many times as necessary to transfer a point of known elevation to another distant point of unknown elevation.

b. Leveling accuracy. Second Order leveling point closure standards for vertical control surveys are shown in Table 3-2. Second Order leveling consists of lines run in only one direction, and between benchmarks previously established by First Order methods. If not checking into another line, the return for Second Order Class I level work should check within the limits of 0.025 times the square root of M feet (where "M" is the length of the level line in miles), while for Second Order Class II work, it should check within the limits of 0.035 times the square root of M feet.

c. Leveling equipment. The type of equipment needed is dependent on the accuracy requirements.

(1) Second Order level. Instruments used in Second Order leveling can be a total station, precise level, or equivalent. Often a graduated parallel plate micrometer is built into the instrument to allow reading to the nearest 0.001 of a unit. The sensitivity of the level vial, telescopic power, focusing distance, and size of the objective lens are factors in determining the precision of the instrument. Instruments are rated and tested according to their ability to maintain the specified order of accuracy. Only those rated as precise geodetic quality instruments may be used for Second Order work.

(2) Precise level rods. Precise level rods are required when running Second Order levels. The rods must be of one piece, invar strip type, with the least graduation on the invar strip of 1 centimeter. The invar strip is 25 millimeters wide and 1 millimeter thick, and is mounted in a shallow groove in a single piece of well-seasoned wood. The front of the rod is graduated in meters, decimeters and centimeters on the invar strip. The back of the rod is graduated in feet and tenths of feet, or yards and tenths of yards. These rods must be standardized by the National Institute of Standards and Technology and their index and length corrections determined. Rods with similar characteristics are paired and marked. The pairings must be maintained throughout a line of levels. The invar strips should be checked periodically against a standard to determine any changes that may affect their accuracy. The precise level rod is a scientific instrument and must be treated as such; not only during use but also during storage and transporting. When not in use they must be stored in their shipping containers to avoid damage. The footpiece should be inspected frequently to make sure it has not been bent or otherwise damaged.

d. Calibrations and adjustments. To maintain the required accuracy, certain tests and adjustments must be made at prescribed intervals to both the levels and rods being used.

(1) Determination of stadia constant. The stadia constant factor of the leveling instrument should be determined by calibration. The stadia factor is required for measurement and computation of distances from the instrument to the leveling rod. This determination is made independently for each level used in the field and is permanently recorded and kept with project files. The determination is made by comparing the measured stadia distance to known distances on a test course. The test course should be laid out on a reasonably level ground and marked with temporary points placed in a straight line at measured distances of 0, 25, 35, 45, 55, 65, and 75 meters. The optical and mechanical centers of the instrument are not necessarily at the same point for a given instrument. Therefore, when determining the stadia constant, this constant should be applied to the measurement before making the test comparison. This constant offset value should be available from the manufacturer's manual. For the test data collection, read the rod at each of the six test points and record the rod intervals. The level bubble should be accurately centered. Each half-wire intervals should be recorded as a check against erroneous readings. The sum of the total intervals for the six readings should be computed. The stadia constant is the sum of these measured distances (300 meters) divided by the sum of the six total wire intervals. As a check against gross errors each separate observation should be computed. The average of the six separate computations serves as a numerical check on the computations.

(2) Determination of "C" Factor. Each day, just before the leveling is begun, or just after the beginning of the day's observations, and immediately following any instance in which the level is subjected to unusual shock, the error of the level, or "C" factor, must be determined. This determination can be made during the regular course of leveling or over a special test course; in either case the recording of the observations must be done on a separate page of the recording notes with all computations shown. If the determination is made during the first setup of the regular course of levels, the following procedure is used (Figure 6-3). After the regular observations at the instrument station "A" are completed, transcribe the last FS reading "a" as part of the error determination; call up the backsight rodman and have the rod placed about 10 meters from the instrument; read the rod "b", over the instrument to a position "B" about 10 meters behind the front rod; read the front rod "c" and then the back rod "d". The two instrument stations must be between the rod points. The readings must be made with the level bubble carefully centered and then all three wires are read for each rod reading. The required "C" factor determined is the ratio of the required rod reading correction to the corresponding subtended interval, or:

$$C = (R1 - R2) / (R3 - R4)$$

where

    R1 = Sum near rod readings
    R2 = Sum distant rod readings
    R3 = Sum distant rod readings
    R4 = Sum near rod readings

The total correction for curvature and refraction must be applied to each distant rod reading before using them in the above formula. It must be remembered that the sum of the rod intervals must be multiplied by the stadia constant in order to obtain the actual distance before correction. The maximum permissible "C" factor varies with the stadia constant of the instrument. The instruments must be adjusted if the "C" factor is:

C	>	0.004	for a stadia constant of 1/100
C	>	0.007	for a stadia constant of 1/200
C	>	0.010	for a stadia constant of 1/333.

The determination of the "C" factor should be made under the expected conditions of the survey as to length of sight, character of ground, and elevation of line of sight above the ground. The date and time must be recorded for each "C" factor determination, since this information is needed to compute leveling corrections.

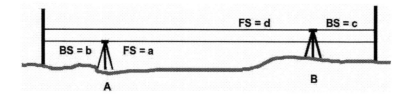

Figure 6-3. C-factor Calibration Procedure

(3) Adjustment of level. The type of instrument being used will dictate the method and procedure used to adjust the instrument if the "C" factor exceeds the allowable limits. The manufacturer's procedures should be followed when adjusting a level.

(4) Test of rod levels. Precise rod levels must be tested once each week during regular use or whenever they receive a severe shock. This test is made with the level rod bubble held at its center, and the deviation of the face and edge of the rod from the vertical are determined. If the deviation from the vertical exceeds 0.01 meter on a 3-meter length of rod, the rod level must be adjusted. The rod level is adjusted in the same manner as any other circular bubble. A statement must be inserted in the records showing the manner in which the test was made, the error that was found, if any, and whether an adjustment was made. When using other than precise leveling rods, this test is not required.

    e. Leveling monumentation. All benchmarks used to monument Second Order level lines will conform to criteria published in EM 1110-1-1002. Benchmarks used to monument Second Order level lines shall be standard USACE brass caps set in concrete. The concrete should be places in holes deep enough to avoid local disturbance. If the brass cap is not attached to an iron pipe, use some type of metal to reinforce the concrete prior to embedding the brass cap. Concrete should be placed in a protected position. If possible, benchmarks should be set close to a fence line, yet far enough away to permit plumbing of level rod. Do not set monuments closer than four feet to a fence post, as the benchmark likely will be disturbed if the post is replaced. Each brass cap must be stamped to identify it by the methods detailed in EM 1110-1-1002. In addition to stamping a local number or name on the cap, it is optional to stamp the elevation on the brass cap after final elevation adjustment has been made. The benchmarks must be set no less than 24 hours in advance of the level crew if the survey is to be made immediately after monument construction.

f. Leveling notes. Notes for Second Order levels will be kept in a manner approved by the survey supervisor. A set style cannot be developed due to different types of equipment that may be employed. Elevations will not be carried in the field as they will be adjusted by the field office and closures approved prior to assigning a final adjusted elevation.

6-3. Third Order Leveling

a. General. Leveling run for traverse profiles, temporary benchmarks, control of cross-sections, slope stakes, soundings, topographic mapping, structure layout, miscellaneous construction layout, and construction staking shall be Third- or Fourth-Order leveling, as detailed in Table 3-2, unless otherwise directed. All levels will originate from and tie into existing control. No level line shall be left unconnected to control unless by specific instructions of the survey supervisor or written directive.

b. Leveling accuracy. All accuracy requirements for USACE vertical control surveys will conform to the point closure standards shown in Table 3-2. The required accuracy for Third Order levels is 0.050 M feet where "M" is the length of the level line in miles, while Construction Layout level work will conform to 0.100 times the square root of M feet. The length of the line may be determined from quad sheets or larger scale map if a direct measure between points is not available.

c. Leveling equipment. The type of equipment needed is dependent upon the accuracy requirements.

(1) Third Order level. A semi-precise level can be used for Third Order leveling, such as the tilting Dumpy type, three-wire reticule, or equivalent.

(2) Leveling rods. The rods should be graduated in feet, tenths and hundreds of feet. The Philadelphia rod or its equivalent is acceptable. However, the project specifications will sometimes require that semi-precise rods be used that are graduated on the front in centimeters and on the back in half foot intervals. The Zeiss stadia rod, fold type, or its equivalent should be used when the specifications require semi-precise rods.

(3) Lower order. The type of spirit level instrument used should ensure accuracy in keeping with required control point accuracy. Precision levels are not required on lower order leveling work. Fennel tilting level, dumpy level, Wye level, or their equivalent are examples of levels that can be used. A stadia rod with least readings of five-hundredths of a foot or 1 centimeter will be satisfactory. The use of turning pins and/or plates will depend upon the type of terrain or if rods may be placed on firm stones or roadways.

d. Leveling monumentation. The level line shall be tied to all existing benchmarks along or adjacent to the line section being run. In the event there are no existing benchmarks near the survey, new ones should be set, not more than 0.5 mile apart. Steep landscape in the area of survey may require monuments to be set at closer spacings.

(1) Benchmarks should be set on permanent structures, such as, head walls, bridge abutments, pipes, etc. Large spikes driven into the base of trees, telephone poles, and fence posts are acceptable for this level of work. All temporary benchmarks must have a full description including location. Unless they are on a turn, they are not considered to be temporary benchmarks. No closures shown by an intermediate shot will be accepted. All temporary benchmarks must have a name or number for future identification.

(2) Turning pins. Turning pins should be driven into in the ground until rigid with no possibility of movement. Turning points or temporary benchmarks will have a definite high point so that any person not familiar with the point will automatically hold the rod on the highest point, and so that it can spin free If solid rocks are being used for turns they must be marked with crayon or paint prior to taking readings.

(3) Rod targets. It is not mandatory to use targets on the rod when the reading is clearly visible. However, they are required in dense brush, when using grade rods, or when unusually long shots are necessary.

e. Leveling notes. Complete notations or sketches will be made to identify level lines and side shots. All Third Order or lower level notes will be completely reduced in the field as the levels are run, with the error of closure noted at all tie in points. In practice, the circuit will be corrected to true at each tie in point unless instructed to do otherwise by the survey supervisor or written directive. Any change in rod reading shall be initialed and dated so there is no doubt as to when a correction was made. Cross out erroneous readings--never erase them. The instrument man shall take care to keep peg notes on all turns in the standard field book. The notes will be dated and noted as to what line is being run, station occupied, identification of turns, etc.

6-4. Mandatory Requirements

Level C-factor (or peg tests) described in this chapter are mandatory for all vertical control surveys.

Chapter 7
Miscellaneous Field Notekeeping and Procedural Requirements

7-1. Field Notes

All field notes will be recorded in a standard hard-cover field book as the measurements are made in the field. The typical dimensions of such a field book are 4-7/8" by 7-1/2".

a. Entries. All field note entries shall be made with a black lead pencil or ink. Notations made by other than the original surveyor shall be made with a colored pencil so a clear distinction exists between the field observations and subsequent corrections, adjustments, comments or supplemental data.

b. Index. The first two pages of each field book shall be reserved for the index and shall not be numbered. The index should contain the date and description of the survey, the type of survey, and the page numbers containing the survey data. The remainder of the field book shall contain the actual field data and shall be numbered beginning at page one.

c. Page setup. The first page of each entry should contain (at the top left side of the page) the name of the installation or project location, a specific project title, and the type of work being done. At the top of the right side of the right half of the page, record the actual date of the survey, weather conditions, type and serial number of instruments used, members of the crew and their assignment, map or field book references, and other remarks as necessary for a complete understanding of the survey.

d. Corrections. No erasures should be made in the field book. If errors are made, they will be crossed through and the correct ones will be written in such a way as the original data remains legible. No figure should be written over the top of another-- nor should any figure be erased. If a whole page is in error, the complete page will be lined or crossed through and the word "VOID" will be written in large letters diagonally across the page. An explanation of the error, and a cross reference will be entered on the voided page showing the book and page number where the correct information may be found. At the end of each day of work, the field notes shall be signed and dated by the individual responsible for the work.

e. Data collector. If a data collector is used, only setup information (station description, HI, sketch, etc.) should be recorded in the field book. This information is used to document the sequence of the survey. Refer to EM 1110-1-1005 for further information on data collector requirements.

f. References. When it is necessary to copy information from another field book or other source, a note will be made which clearly states that the information was copied and the source from which it came. If the notes are a continuation from another field book, a description will be written in the field book to the effect "NOTES CONTINUED FROM BK XXXX PAGE XX". A similar description (e.g., CONTINUED IN BOOK XXXX FROM PAGE XX) will be written on the last page of each section of notes if those notes are to be continued either in another book or on another page which is not adjacent to the current page.

g. Sketches. The sketch should show all the details, dimensions, and explanatory notes required. The sketch should be written on a whole page whenever possible. If necessary, multiple pages with the sketch divided equally among the pages should be used if the sketch has too many details to be shown on one page.

7-2. Horizontal Control Survey Field Notes

a. Data entries. Traverse field notes shall contain for each occupation: the height of the instrument above the station occupied, the target height above the stations measured to, both horizontal and vertical angles, and either slope or horizontal distance.

b. Station description. A description of the point occupied shall be made in the field notes. This description shall include the type of monument, its general location, and the type of material the point is set in. A sketch of the location of the point relative to existing physical features and reference ties shall be made and included in the notes. If a horizontal control line is used, a sketch of it shall be made and included in the notes. This sketch does not need to drawn to scale, but it should include the relative position of one point to the next and the basic control used.

7-3. Vertical Control Survey Field Notes

a. Description. A short description of the course of the level line shall be entered in the field book.

b. References. Entries shall be made in the book that give the references to the traverse notes and other existing data used for elevations (e.g., TRAVERSE BOOK XXXX PAGE XX, USGS Quad XXXXXX, NOS Chart XXXX, etc.).

c. *Benchmark description.* A complete description of each point on which an elevation is established shall be recorded in the field book adjacent to the station designation.

7-4. Rights-of-Entry

a. General. When entering property to conduct a survey, rights of the property owner will be respected. The following details some minimum guidelines to follow.

b. Permission. Permission to enter a military installation and other private property will always be acquired by the District prior to entering such property. While on the military installation, members of the survey crew will adhere to all of the stipulations (e.g., rules, regulations, directives, verbal guidance, etc.) set forth by the Installation Commander or his designated representative. The same basic guidelines are applicable when the right to enter private property is given.

c. Property. Government and private property shall be protected at all times. Every effort should be made not to damage or cut trees, shrubs, plants, etc. on the property. If such must be done, the Installation Commander, or in the case of private property, the private property owner, is the only person who can grant permission to do so. It shall be standard practice that property entered shall be returned to its condition prior to entry once the survey is completed. Gates and other structures should be left in the position in which they were found prior to entry. If a gate is closed, do not leave it open for any long period of time. Return all borrowed property (e.g., keys, maps, etc.) as instructed by the property owner or designated representative.

d. Monuments. Survey points should be placed in such a way as to not obstruct the operations of the property owners or be offensive to their view. Monuments set as a result of the survey should be set below ground level to prevent damage by or to any equipment or vehicles. Extra care must be taken when setting a survey point at or near airports. Any pre-marks set on military installations or private property will be removed as soon as possible after the survey work is completed, or at the request of the Installation Commander, property owner, and/or designated representative.

7-5. Mandatory Requirements

There are no mandatory requirements in this chapter.

Appendix A
References

ER 405-1-12
Real Estate Handbook

ER 1110-2-100
Periodic Inspection and Continuing Evaluation of Completed Civil Works Structures

ER 1110-2-1150
Engineering and Design for Civil Works Projects

EM 1110-1-1000
Photogrammetric Mapping

EM 1110-1-1002
Survey Markers and Monumentation

EM 1110-1-1003
NAVSTAR Global Positioning System Surveying

EM 1110-1-1005
Topographic Surveying

EM 1110-1-2909
Geospatial Data and Systems

EM 1110-2-1003
Hydrographic Surveying

EM 1110-2-1908
Instrumentation of Embankment Dams and Levees

EM 1110-2-4300
Instrumentation for Concrete Structures

USATEC SR-7 1996
"Handbook for Transformation of Datums, Projections, Grids and Common Coordinate Systems"

US Bureau of Land Management 1947
"Manual of Instruction for the Survey of the Public Lands of the United States"

FGCS 1984
Federal Geodetic Control Subcommittee 1984
"Standards and Specifications for Geodetic Control Networks," Rockville, MD.

FGCS 1980
Federal Geodetic Control Subcommittee 1980
"Input Formats and Specifications of the National Geodetic Data Base" (also termed the "Bluebook")

FGCS 1988
Federal Geodetic Control Subcommittee 1988
"Geometric Geodetic Accuracy Standards and Specifications for Using GPS Relative Positioning
Techniques (Preliminary)," Rockville, MD. (Reprinted with Corrections: 1 Aug 1989).

NGS 1988
National Geodetic Survey 1988
"Guidelines for Submitting GPS Relative Positioning Data to the National Geodetic Survey"

Appendix B
CORPSCON Technical Documentation and Operating Instructions

B-1. General

a. Background. The National Geodetic Survey has developed three programs called NADCON (North American Datum Conversion), VERTCON (Vertical Conversion), and Geoid96. NADCON provides consistent results when converting to and from the North American Datum of 1983 (NAD 83) and the North American Vertical Datum of 1988 (NAVD 88). NADCON converts coordinates between NAD 83 and the following datums; NAD 27, Old Hawaiian Datum, Puerto Rico Datum, St. George Island (Alaska) Datum, St. Paul Island (Alaska) Datum and St. Lawrence Island (Alaska) Datum. For organizational purposes, the latter six datums are referred to as NAD 27 within the program. VERTCON converts orthometric heights between National Geodetic Vertical Datum of 1929 (NGVD 29) and NAVD 88. Geoid96 calculates the separation between the Geoid and the Geodetic Reference System of 1980 (GRS 80) ellipsoid. NADCON, VERTCON, and Geoid96 work exclusively in geographical coordinates (Latitude/Longitude).

b. CORPSCON. The US Army Topographic Engineering Center (USATEC) created a more comprehensive program called CORPSCON (Corps Convert), which is based on NADCON, VERTCON and Geoid96. In addition to transformations between NAD 83 and NAD 27 geographical coordinates, CORPSCON also converts between State Plane Coordinates Systems (SPCS), Universal Transverse Mercator (UTM) and geographical coordinates; thus eliminating several steps in the total process of converting between SPCS 27, SPCS 83, UTM 27, and UTM 83. Inputs can be in either geographic or SPCS/UTM coordinates (SPCS 27 X and Y or SPCS 83 Northing and Easting). This program can also be used to convert between state plane, geographic, and UTM coordinates on the same datum. CORPSCON will convert orthometric and ellipsoidal heights in Geographic, State Plane and UTM coordinate systems. CORPSCON allows conversions based on US Survey and International Feet. As of 1997, 19 states have specified, by statute, units of measure for grid coordinates as follows:

(1) US Survey Foot - California, Colorado, Connecticut, Idaho, Indiana, Kentucky, Maryland, Massachusetts, Mississippi, New Mexico, New York, North Carolina, Oklahoma, Pennsylvania, Tennessee, Texas, Washington and Wisconsin.

(2) International Survey Foot - Arizona, Michigan, Montana, North Dakota, Oregon, South Carolina and Utah.

c. Horizontal datums. The Federal Geodetic Control Subcommittee (FGCS) has adopted NAD 83 as the official horizontal datum for US surveying and mapping activities performed or financed by the Federal Government (Federal Register / Vol. 54, No. 113, June 14, 1989). The FGCS also stated that NADCON will be the standard conversion method for all mathematical transformations between NAD 83 and NAD 27. CORPSCON includes conversions based on High Accuracy Regional Networks (HARN).

d. Vertical datums. FGCS has affirmed that NAVD 88 shall be the official vertical reference datum for the United States (Federal Register / Vol. 58, No. 120, June 24, 1993).

e. Coverage. The current version performs NAD 27/NAD 83 and NAVD 88/GRS 80 conversions for the continental US (CONUS), including the 200 mile commercial zone, Alaska, Hawaii, Puerto Rico, and the US Virgin Islands. Current (1996) areas of coverage for HARNs are Alabama, Arizona, California, Colorado, Florida, Idaho-Montana, Kentucky, Louisiana, Maine, Maryland-Delaware, Mississippi, Nebraska, New England (Connecticut, Massachusetts, New Hampshire, Vermont),

New Mexico, Oklahoma, Puerto Rico-Virgin Islands, Tennessee, Texas, Virginia, Washington-Oregon, Wisconsin and Wyoming. The current version performs NGVD 29/NAVD 88 conversions for the continental US only.

f. Accuracy. NADCON and VERTCON transformations between datums are based on a model of over 250,000 common stations. Therefore, conversions are approximate and accuracy can vary depending on location and proximity to common stations. The accuracy of the NADCON transformations should be viewed with some caution. At the 67 percent confidence level, this method introduces approximately 0.15 meter uncertainty within the conterminous United States, 0.50 meter uncertainty within Alaska, 0.20 meter uncertainty within Hawaii and 0.05 meter uncertainty within Puerto Rico and the US Virgin Islands. In areas of sparse geodetic data coverage, NADCON may yield less accurate results, but seldom in excess of 1.0 meter. Transformations between NAD 83 and states/regions with High Accuracy Reference Networks (HARNs) introduce approximately 0.05 meter uncertainty. Transformations between old datums (NAD 27, Old Hawaiian, Puerto Rico, etc.) and HARN could combine uncertainties (i.e., NAD 27 to HARN equals 0.15m + 0.05m = 0.20m). In near offshore regions, results will be less accurate, but seldom in excess of 5.0 meters. Farther offshore NAD 27 undefined. Therefore, the NADCON computed transformations are extrapolations and no accuracy can be stated. The VERTCON 2.0 model was computed on May 5, 1994 using 381,833 datum difference values. A key part of the computation procedure was the development of the predictable, physical components of the differences between the NAVD 88 and NGVD 29 datums. This included models of refraction effects on geodetic leveling, and gravity and elevation influences on the new NAVD 88 datum. Tests of the predictive capability of the physical model show a 2.0 cm RMS agreement at our 381,833 data points. For this reason, the VERTCON 2.0 model can be considered accurate at the 2 cm (one sigma) level. Since 381,833 data values were used to develop the corrections to the physical model, VERTCON 2.0 will display even better overall accuracy than that displayed by the uncorrected physical model. This higher accuracy will be particularly noticeable in the eastern United States.

B-2. Source of Program and Assistance

To obtain copies of the CORPSCON program, contact:

> US Army Topographic Engineering Center
> ATTN: CEERD-TS-G
> 7701 Telegraph Road
> Alexandria, Virginia 22315-3864
> (703) 428-6766
> or
> http://www.tec.army.mil

and follow links to software distribution.

B-3. Hardware and Software Requirements

a. Hardware. A 80486 (or higher) PC with and 20 MB of hard disk space is required. CORPSCON is compatible with most PC monitors, although color monitors (EGA and VGA) provide the most favorable and easily discernible menu display. If a printer is to be used, it must be interfaced through the parallel port (LPT1 or LPT2) in order to be recognized by the program.

b. Software. CORPSCON runs under MS-Windows 3.1, MS-Windows 95 or MS-Windows NT. The CONFIG.SYS file must have FILES set to 25.

B-4. Installation Procedures

a. Diskette installation. To install the CORPSCON program from diskette, perform the following steps:

(1) Insert Distribution Disk #1 in the a: drive of the computer.

(2) For Windows 95, select Run from the Start menu. For Windows 3.1, go to the Program Manager group box and select File and then Run. A Run window should appear.

(3) Enter 'a:setup' in the Command Line item of the Run window. This should activate the CORPSCON installation window.

(4) Follow the directions on screen to install the program.

b. CD installation. To install the CORPSCON program from CD, perform the following steps:

(1) Insert the CD in the computer.

(2) For Windows 95, select Run from the Start menu. For Windows 3.1, go to the Program Manager group box and select File and then Run. A Run window should appear.

(3) Enter 'd:\corpscon\setup' in the Command Line item of the Run window. This should activate the CORPSCON installation window. If the CD-ROM is on a drive other than the >d:= drive, replace the >d:= with the appropriate drive letter.

(4) Follow the directions on screen to install the program.

B-5. Program and Data Files

a. Program Files. Upon installation, the following files should be located in the destination directory (c:\CORPSCON if the default installation was used):

corpswin.exe	corpswin.cfg	utms.hlp
geoareas.lst	corpswin.wri	geoid96.txt
vertcon.txt	conus.las	conus.los
hawaii.las	hawaii.los	prvi.las
prvi.los	vertcone.94	vertconc.94
vertconw.94	geoid96ne.geo	geoid96nc.geo
geoid96nw.geo	geoid96se.geo	geoid96sc.geo
geoid96sw.geo	haw96.geo	prvi96.geo
alhpgn.las	alhpgn.los	azhpgn.las
azhpgn.los	cahpgn.las	cahpgn.los
cohpgn.las	cohpgn.los	emhpgn.las
emhpgn.los	ethpgn.las	ethpgn.los
flhpgn.las	flhpgn.los	kyhpgn.las
kyhpgn.los	lahpgn.las	lahpgn.los
mehpgn.las	mehpgn.los	mdhpgn.las
mdhpgn.los	mshpgn.las	mshpgn.los
nehpgn.las	nehpgn.los	nmhpgn.las
nmhpgn.los	okhpgn.las	okhpgn.los

pvhpgn.las	pvhpgn.los	tnhpgn.las
tnhpgn.los	vahpgn.las	vahpgn.los
wihpgn.las	wihpgn.los	wmhpgn.las
wmhpgn.los	wohpgn.las	wohpgn.los

If the Alaska Data Files were installed, the following files should be copied to the destination directory (c:\CORPSCON if the default installation was used):

alaska.las	alaska.los	stlrnc.las
stlrnc.los	stgeorge.las	stgeorge.los
stpaul.las	stpaul.los	geo96an.geo
geo96as.geo		

b. LAS/.LOS, .94, and .GEO Files. Files with .las and .los extensions are NADCON data files. These files are used for NAD 27/NAD 83/HPGN conversions. Files with .94 extensions are VERTCON data files. These files are used for NGVD 29/NAVD 88 conversions. File with .geo extensions are Geoid96 files. These files are used for GRS 80/NAVD 88 conversions.

c. CORPSCON.CFG File. When CORPSCON is run, the CORPSCON.cfg file will be updated. This file will hold all of the configuration information for the most recent conversion. Information maintained includes input and output datums, zones, units, and output file names.

d. CORPSCON.INI File. The CORPSCON.ini file will be created in the Windows directory by the installation program. The CORPSCON.ini file contains several variables required for program execution. Each variable holds a directory name as specified below.

    programfiles - directory for all program files (corpswin.exe & utms.hlp)

    nadconfiles - directory for all NADCON (.las & .los) files

    vertconfiles - directory for all VERTCON (.94) files

    geoid9396files - directory for all Geoid96 (.geo) files

    tempfiles - directory where temporary files are created

    configfiles - directory where the configuration file (corpswin.cfg) is stored

The CORPSCON.ini file also contains descriptions and base filenames of HARN areas. This file may be modified to include new or updated HARN files. The format of entries in this file is:

    description = basefilename

For example, the files used to cover Maryland and Delaware are mdhpgn.las and mdhpgn.los. The corresponding entry in the CORPSCON.ini file would be:

    Maryland - Delaware = mdhpgn

The .las and .los file extensions should NOT be included in the filename. The CORPSCON.ini file may be modified by hand using any text editor, or entries may be added by using the Utilities/Add New HPGN File option.

e. Geoareas.lst file. The geoareas.lst file contains a list of all Geoid96 data files. This file must be present or CORPSCON will default to using Geoid93 data files.

B-6. Operating Instructions

a. General. Execute the CORPSCON program through the Start Menu for Windows 95 or through the CORPSCON icon for Windows 3.1. This should open the CORPSCON Main Window. The CORPSCON Window consists of four items: the Main Menu Bar, the Input Format information box, the Output Format information box, and the Send Data information box. All user interaction is performed through the use of the main menu bar. The information boxes are included for reference purposes only. The basic procedure for performing a conversion is:

(1) Specify input data information using the Input Data Format menu item.

(2) Specify output data information using the Output Data Format menu item.

(3) Specify the devices/files to which the data should be sent using the Send Data menu item.

(4) Perform the conversion using the Convert menu item.

The main menu bar consists of six menu items: Convert, Input Format, Output Format, Send Data, Utilities, and Help.

b. Convert. The convert menu item has three sub-items:

- Single Point (Manual Input)
- CORPSCON Batch File
- User Defined Input File

(1) Convert/Single Point. The single point sub-item is used to convert a single data point. When this item is selected, a window will appear prompting the user to input relevant information. For grid coordinate conversions, the user must enter the Northing and Easting or X and Y values. For geographic conversions, the user will need to enter the latitude and longitude. If vertical conversions are being performed, the user must also enter an elevation value. An optional point name may be entered. The user should enter in the appropriate information and press the OK button to perform the conversion. If data is to be sent to an output and/or printer file, other windows will appear which will allow the specification of the names of these output files. If data is to be sent to an Output Window, the results of the conversion will appear in a separate window.

(2) Convert/CORPSCON Batch File. The CORPSCON Batch File sub-item is used to convert files, which are in the standard CORPSCON Batch File Format. Files in this format may be created by using the Utilities/Build New CORPSCON Batch File menu item. The details of this file format are included below. When this sub-item is selected, a window will appear prompting the user to select the name of the input CORPSCON Batch File. The user should select an input filename and press the OK button to perform the conversion. If data is to be sent to an output and/or printer file, other windows will appear which will allow the specification of the names of these output files. If data is to be sent to an Output Window, the results of the conversion will appear in a separate window.

(3) CORPSCON Batch File Format. A CORPSCON Batch File is an ASCII text file containing three or four comma-delimited fields. For geographic coordinates the fields are:

Point Name
Latitude
Longitude
Elevation (optional)

Latitude and longitude values may be in decimal degrees, degrees-decimal minutes, or degrees-minutes-decimal seconds. Longitude values have a positive west sign convention. Degree, minute, and second values must be separated by a space within the latitude or longitude field. The point name is not required, but a ',' must appear before the latitude value in order to be accepted as a valid line. The fourth field is required only if vertical conversions are to be performed. For grid coordinates, the fields are:

Point Name
Easting or X value
Northing or Y value
Elevation (optional)

Again, the fourth field is required only if vertical conversions are to be performed. Lines beginning with a ' ; ' or ' # ' characters in CORPSCON Batch Files are interpreted as comment lines. No conversion of data will be performed for comment lines.

    c. Convert/User defined input file. The User Defined Input File sub-item is used to convert files in a format which is specified by the user. These files may contain up to six fields and may be comma or space delimited. When this option is selected, a window will appear prompting the user to specify the format of the input file. At a minimum, the type of data each field will contain and the delimiter of fields (comma or space) in the data file must be specified. If the input file contains geographic coordinates, the format of these coordinates must also be specified. Geographic coordinates may be in decimal degrees, degrees-decimal minutes or degrees-minutes-decimal seconds. The user should specify the format of the input file and press the OK button to continue. The User Defined File window and examples of its use are included below. After specification of the input file format, a window will appear which will allow entry the input data file name. The user should enter the name and press the OK button to continue. If data is to be sent to an output file, another User Defined File window will appear which allows specification of the format of the output file. The output file may have a different format than that of the input file. The user should specify the format of the output file and press the OK button to continue. After specification of the output file format, a window will appear which will allow entry of the output data file name. The user should enter the name and press the OK button to perform the conversion. If data is to be sent to a printer file, another window will appear which will allow specification of the name of the printer file. If data is to be sent to an Output Window, the results of the conversion will appear in a separate window.

    (1) User Defined File Dialog Box. The User Defined File dialog box consists of two blocks of information: the field specifications and other file information. The six fields that a user defined file may have are:

Point Name,
Northing/Y/Latitude,
Easting/X/Longitude,
Elevation,
Carry Field 1,
Carry Field 2.

The Carry Fields act as place-holders of extra information which may be included in the file but is not necessary for the conversion. These carry fields can be included in an output file. The field delimiter of the file must also be specified by selecting commas or spaces in the Delimiter drop-down box. If the input file contains geographic coordinate information, the coordinate format must also be specified in the Degree Format drop down box. Valid formats are decimal degrees, degrees-decimal minutes or degrees-minutes-decimal seconds. Geographic coordinates that are in degrees-decimal minutes or degrees-minutes-decimal seconds must have a space between the degree-minute and minute-second values in the input file. If the user-defined file is to be an output file, header information may be included in the output. Header information will contain data on the output datums, data/time, and company/project names. These lines will have a ';' as their leading character which indicates a comment.

d. Input/Output Format Menu Items. Input/Output Format menu items allow the user to select Geographic, UTM, or State Plane coordinates on NAD 27, NAD 83, or HPGN.

(1) Geographic Coordinates. Geographic coordinates are selected, a window will appear which allows the user to specify the vertical datum (NGVD 29, NAVD 88 or GRS 80) and units.

(2) UTM Coordinates. UTM coordinates are selected a window will appear which prompts the user to input the UTM zone, horizontal units, vertical datum (if any) and vertical units.

(3) SPCS Coordinates. State Plane coordinates are selected, a window will appear which prompts the user to input the State Plane zone, horizontal units, vertical datum (if any) and vertical units.

(4) HPGN Coordinates. HPGN conversion is selected a second dialog box will appear prompting the user to selected the desired area for the HPGN conversion.

A check will appear next to the currently selected format.

e. Send Data. The send data menu items allow the user to specify where the output data should be sent. Data may be sent to an Output Window, Output File, Printer File or to the Printer. Data may be sent to more than one device or file. A check will appear next to the sub-item that is to receive output data.

f. Utilities. The Utilities menu item has the following sub-items:

Build CORPSCON Batch File
Append Existing CORPSCON Batch File
View NADCON/VERTCON/Geoid96 File Status
Degree Conversion and Preferences.

(1) Build CORPSCON Batch File. This sub-item allows the user to create input files in the CORPSCON Batch File Format. The user may build geographic or grid files which may or may not contain elevation values.

(2) Append Existing Batch File. This sub-item allows the user to add points to an existing CORPSCON Batch File.

(3) View NADCON/VERTCON/Geoid93/96 File Status. CORPSCON requires that several data files be located and opened successfully in order to perform NAD 27/NAD 83 and NGVD 29/NAVD 88/GRS 80 conversions. This sub-item allows the user to determine which data files have been opened successfully.

(4) Degree Conversion. This tool allows the user to input a degree value in decimal degrees, degrees-decimal minutes, or degrees-minutes-decimal seconds. The value is then displayed in an output window in decimal degrees, degrees-decimal minutes, and degrees-minutes-decimal seconds.

(5) Add New HPGN File. This tool allows the user to add new or updated HPGN files to the program. The user should enter the area description and the base filename. The base filename should NOT include .las or .los extensions.

(6) Preferences. The Preferences dialog box allows specification of a Company and Project name. It also allows specification of the NADCON, VERTCON, and Geoid96 files directories. Grid coordinate entry and display may also be set here. A Northing-Easting or X-Y display may be selected.

g. Help. The Help menu item has the following sub-items:

(1) UTM Zones. This sub-item will display information about UTM zones. A diagram displaying approximate UTM zones for the continental US is included.

(2) About. The About sub-item will display information about the program including version number.

B-7. Error Messages

a. General. CORPSCON is designed to prompt the user for most cases in which a system or runtime error occurs. These errors are as follows:

(1) No Math Co-Processor. A Math co-processor (hardware item) must be installed in the computer to run the program.

(2) No Input File. A file with the name specified was not found; enter the correct file name or create the file before running CORPSCON. Check to be certain that the file and program CORPSCON are in same directory and sub-directory.

(3) Printer Error. The program is unable to send the output to the printer. Check the following:

printer connection
printer turned on
printer interfaced through the parallel port
paper in printer
printer in "on-line" or "ready-to-print" status

Appendix C
Development and Implementation of NAVD 88

C-1. General Background

This appendix provides technical guidance and implementation procedures for the conversion from the National Geodetic Vertical Datum of 1929 (NGVD 29) to the North American Vertical Datum of 1988 (NAVD 88).

a. The NAVD 88 is a new vertical datum for North America that effectively covers Canada, Mexico, and the US. The new adjustment of the US National Vertical Control Network (NVCN) was authorized in 1978, and in 1982 the National Oceanic and Atmospheric Administration (NOAA) and Canada signed a Memorandum of Understanding (MOU) regarding the adoption of a common, international vertical control network called the NAVD 88.

b. The Federal Geodetic Control Subcommittee (FGCS) of the Federal Geographic Data Committee (FGDC) has adopted the new NAVD 88 datum. In addition, NAVD 88 was established in conjunction with the International Coordinating Committee on Great Lakes Basic Hydraulic and Hydrologic Data. This committee defined the IGLD 85, which was published for use in January 1992. IGLD 85 replaced IGLD 55.

C-2. References

a. Water Resources Development Act of 1992 (WRDA 92), Section 224, Channel Depths and Dimensions.

b. EM 1110-2-1003, Hydrographic Surveying.

c. Converting the National Flood Insurance Program to the North American Vertical Datum of 1988: Guidelines for Community Officials, Engineers, and Surveyors, FEMA Report No. FIA-20, June 1992.

d. Results of the General Adjustment of the North American Datum of 1988, American Congress on Surveying and Mapping Journal of Surveying and Land Information Systems, Vol. 52, No. 3, 1992, pp. 133-149.

e. American Congress on Surveying and Mapping Ad Hoc Committee Report on NAVD 88, Special ACSM Report, 1990.

C-3. Discussion

a. NGVD 29 has been replaced by NAVD 88, an international datum adopted for use in Canada, the United States and Mexico. NAVD 88 was established to resolve problems and discrepancies in NGVD 29. The adjustment of NAVD 88 was completed in June 1991 by the National Geodetic Survey (NGS), an agency of the Department of Commerce, National Oceanic and Atmospheric Administration (NOAA). NAVD 88 was constrained by holding fixed the height of a single primary tidal benchmark (BM) at Father's Point/Rimouski, Québec, Canada, and performing a minimally constrained general adjustment of US-Canadian-Mexican leveling observations. The result of this adjustment is newly published NAVD 88 elevation values for benchmarks (BMs) in the NGS inventory. Most Third Order BMs, including those of other Federal, state and local government agencies, were not included in the

NAVD 88 adjustment. Appendix A contains further background information on the development and adjustment of NAVD 88.

b. The Federal Geodetic Control Subcommittee (FGCS) of the Federal Geographic Data Committee (FGDC) has affirmed that NAVD 88 shall be the official vertical reference datum for the US. The FGDC has prescribed that all surveying and mapping activities performed or financed by the Federal Government make every effort to begin an orderly transition to NAVD 88, where practicable and feasible

c. Both tidal and non-tidal low water reference planes and datums are affected by the change to NAVD 88. The datum for the Great Lakes is now the International Great Lakes Datum of 1985 (IGLD 85). Unlike the prior datum (IGLD 55), IGLD 85 has been directly referenced to NAVD 88 and originates at the same point as NAVD 88. Elevations of reference points/datums along the various inland waterway systems will also be impacted by the change in datums.

d. The transition to NAVD 88 may have considerable impact on Corps projects, including maps, drawings, and other spatial data products representing those projects. However, once completed, the transition will result in a more accurate vertical reference datum that has removed leveling errors, accounts for subsidence, and other changes in elevation.

e. The computer program VERTCON can be used to make approximate conversions between NGVD 29 and NAVD 88. This program was developed by NGS and during the later part of FY94 has been incorporated into the USACE program CORPSCON. VERTCON conversions are intended for general small-scale mapping uses -- VERTCON should not be used for converting benchmark elevations used for site plan design or construction applications.

f. The conversion to NAVD 88 should be accomplished on a project by project basis. The relationship of all project datums to both NGVD 29 and NAVD 88 will be clearly noted on all drawings, charts, maps, and elevation data files.

g. In accordance with Section 24 of WRDA 92, when elevations are referred to a tidal reference plane in coastal waters of the US, Mean Lower Low Water (MLLW) shall be used as the vertical datum-- see Appendix D in this manual. Tidal BMs should be tied to NAVD 88 instead of NGVD 29 where NAVD 88 data is available. Tidal datums shall be established in accordance with the procedures outlined in EM 1110-2-1003.

h. Other hydraulic-based reference planes established by USACE for the various inland waterways, reservoirs, and pools between control structures should continue to be used for consistency; however, they should also be connected with the NAVD 88 where practicable and feasible.

i. In project areas where local municipal or sanitary jurisdictions have established their own vertical reference planes, every attempt should be made to obtain the relationship between that local datum and NGVD 29 and/or NAVD 88; and clearly note this relationship on all drawings, charts, maps, and elevation data files.

C-4. The National Vertical Control Network (NVCN): NGVD 29 and NAVD 88

a. The NVCN consists of a hierarchy of interrelated nets that span the United States. Before the adoption of NAVD 88, benchmark (BM) elevations of the NVCN were published as orthometric heights referenced to NGVD 29. NGVD 29 was established by the United States Coast and Geodetic Survey (USC&GS) 1929 General Adjustment. NGVD 29 was established by constraining the combined US and Canadian first order leveling nets to conform to Mean Sea Level (MSL) as determined at 26 long term tidal gage stations that were spaced along the east and west coast of North American and along the Gulf of Mexico, with 21 stations in the US and 5 stations in Canada.

b. Local MSL is a vertical datum of reference that is based upon the observations from one or more tidal gaging stations. NGVD 29 was based upon the assumption that local MSL at those 21 tidal stations in the US and 5 tidal stations in Canada equaled 0.0000 foot on NGVD 29. The value of MSL as measured over the Metonic cycle of 19 years shows that this assumption is not entirely valid and that MSL varies from station to station.

c. The NGVD 29 was originally named the Mean Sea Level Datum of 1929. It was known at the time that because of the variation of ocean currents, prevailing winds, barometric pressures and other physical causes, the MSL determinations at the tide gages would not define a single equipotential surface. The name of the datum was changed from the Mean Sea Level Datum to the NGVD 29 in 1973 to eliminate the reference to sea level in the title. This was a change in name only--the definition of the datum established in 1929 was not changed. Since NGVD 29 was established, it has become obvious that the geoid based upon local mean tidal observations would change with each measurement cycle. Estimating the geoid based upon the constantly changing tides does not provide the most stable estimate of the shape of the geoid, or the basic shape of the earth.

d. The datum for NAVD 88 is based upon the mass or density of the Earth instead of the varying heights of the seas. Measurements in the acceleration of gravity are made at observation points in the network and only one datum point, at Pointe-au-Pere/Rimouski, Québec, Canada, is used. The vertical reference surface is therefore defined by the surface on which the gravity values are equal to the control point value. Although the international cooperation between the United States and Canada greatly strengthened the 1929 network, Canada did not adopt the 1929 vertical datum. The NGVD 29 was strictly a national datum. NAVD 88 is an international vertical datum for the US, Canada, and Mexico.

C-5. Distinction Between Orthometric and Dynamic Heights

a. There are several different reference elevation systems used by the surveying and mapping community. Two of these height systems are relevant to IGLD 85: orthometric heights and dynamic heights. Geopotential numbers relate these two systems to each other. The geopotential number (C) of a BM is the difference in potential measured from the reference geopotential surface to the equipotential surface passing through the survey mark. In other words, it is the amount of work required to raise a unit mass of 1 kg against gravity through the orthometric height to the mark. Geopotential differences are differences in potential which indicate hydraulic head. The orthometric height of a mark is the distance from the reference surface to the mark, measured along the line perpendicular to every equipotential surface in between. A series of equipotential surfaces can be used to represent the gravity field. One of these surfaces is specified as the reference system from which orthometric heights are measured. These surfaces defined by the gravity field are not parallel surfaces because of the rotation of the earth and gravity anomalies in the gravity field. Two points, therefore, could have the same potential but may have two different orthometric heights. The value of orthometric height at a point depends on all the equipotential surfaces beneath that point.

b. The orthometric height (H) and the geopotential number (C) are related through the following equation:

$$C = G \cdot H$$

(Eq C-1)

where G is the gravity value estimated for a particular system. Height systems are called different names depending on the gravity value (G) selected. When G is computed using the Helmert height reduction formula that is used for NAVD 88, the heights are called Helmert Orthometric Heights. When G is computed using the International Formula for Normal Gravity, the heights are called Normal Orthometric Heights. When G is equal to normal gravity at 45 deg latitude, the heights are called Normal Dynamic Heights. It should be noted that dynamic heights are just geopotential numbers scaled by a constant, using normal gravity at 45 deg latitude equal to 980.6199 gals. Therefore, dynamic heights are also an estimate of hydraulic head. In other words, two points that have the same geopotential number will have the same dynamic height.

c. IGLD 55 is a normal dynamic height system that used a computed value of gravity based on the International Formula for Normal Gravity. Today, there is sufficient observed gravity data available to estimate "true" geopotential differences instead of "normal" geopotential differences. The "true" geopotential differences, which were used in developing IGLD 85 and NAVD 88, will more accurately estimate hydraulic head.

C-6. Problems with NGVD 29 and Why a New Datum Needed to be Established

a. Approximately 625,000 km of leveling have been added to the National Geodetic Reference System (NGRS) since the 1929 adjustment. Each new line has been adjusted to the network. The new leveling data uncovered some problems in NGVD 29. Through the years, the agreement between the new leveling and the network BM elevations slowly grew worse. An investigation of NGVD 29 general adjustment results indicates that large residuals were distributed in some areas of the country during that adjustment. For example, the accumulated 1929 adjustment correction along a 3000 km east-west leveling route from Crookston, Minnesota, to Seattle, Washington is a delta of 89 centimeters (cm).

b. Inconsistencies in NGVD 29 have increased over the years. This increase is a function of factors such as:

(1) Many BMs were affected by unknown vertical movement due to earthquake activity, post-glacial rebound, and ground subsidence.

(2) Numerous BMs were disturbed or destroyed by highway maintenance, building, and other construction.

(3) New leveling became more accurate because of better instruments, procedures, and improved computations. It was decided in 1977 that the high accuracy achieved by the new leveling was being lost when forced to fit the 1929 network, and plans were made to begin developing a new national vertical network.

c. These inconsistencies have not always been apparent to the user since NGS has periodically readjusted large portions of the NVCN and spread these large residuals over large areas. Eventually, however, there would be a large number of areas in which surveyors would not be able to check their work using NGVD 29. NAVD 88 is specifically designed to remove the inconsistencies and distortions such as those found in the NGVD 29. NGS has held off incorporating approximately 40,000 km of newer leveling data for these reasons. These data were incorporated into the NAVD 88.

C-7. Selection of the Adjustment Method for NAVD 88

a. The FGCS created a Vertical Subcommittee in 1989 to study the impact of the NAVD 88 on the programs of member agencies and to recommend a datum definition. Several different datum definitions for NAVD 88 were studied by the subcommittee and the three options below were selected for final consideration:

(1) Fix the elevation or mean sea level at a single point.

(2) Fix mean sea level at four tide gages located at the network corners.

(3) Fix the NGVD 29 elevations at 18 existing, well scattered BMs.

b. Two options were considered for the establishment of the vertical datum: (1) tidal epoch or (2) a minimally constrained adjustment. The tidal epoch option required that the adjustment hold MSL fixed at all appropriate primary tide stations and use the latest available tidal epoch. This definition is actually the same as NGVD 29, but used the latest data available. The other option used a minimally constrained adjustment holding local MSL fixed at one primary tide station and adjusting all leveling data to it. This second option would maintain the integrity of the leveling data but would also create the greatest deviation from the presently published data.

c. Research was done by NGS to determine which option would be the best choice. To assist in the NAVD 88 datum definition decision, several adjustments were performed using different constraints. The results obtained from several trial adjustments indicate that no matter which datum definition scenario is chosen for NAVD 88, including a minimally constrained adjustment, that changes in absolute heights of as much as 75 to 100 cm would exist between NGVD 29 and NAVD 88.

d. In addition to the NGS research, agencies and appropriate bodies were queried to determine which option would be their preference and to ask for recommendations. The FGCS and the American Congress on Surveying and Mapping (ACSM) established committees to investigate the impact of NAVD 88 on their members' activities and the activities of others in the community. Members of these committees were requested to provide documentation on the affects that the readjustment would have on their user populations and to include specific examples describing the real impact of a new vertical datum on their products. USACE was included in the questionnaire survey. The ACSM report is included as a reference to this document.

e. As a preliminary measure, both committees drafted recommendations for NAVD 88 and specified that NGS should:

(1) Perform minimally-constrained least squares adjustment of the data for NAVD 88.

(2) Shift the datum vertically to minimize recompilation of National Mapping Products.

(3) Develop computer transformation software to convert between NGVD 29 and NAVD 88. ("VERTCON").

(4) Develop national and/or regional geoid models to ensure GPS height differences meet at least 2nd Order, Class II FGCS precise geodetic leveling standards for accuracy.

f. Results indicated that the tidal epoch option would minimize the magnitude of the changes from NGVD 29 to NAVD 88 and thus possibly allow direct comparison of present hydrographic survey elevations with the proposed new NAVD 88 elevations. The smaller change between elevations would cause less confusion and concern over flow heights, and the like. Regardless of the datum definition selected, large differences would exist between the NAVD 88 and the NGVD 29 heights. It should be noted that the NAVD 88 heights are better estimates of orthometric heights than the NGVD 29 heights.

Better estimates of orthometric heights will become more critical in the future as surveying techniques continues to become more sophisticated and more accurate. The improved accuracy of geoid height determinations using GPS data requires the best estimate of true orthometric heights. Many cartographers want heights on their maps based on the best estimate of true orthometric heights.

C-8. The NAVD 88 Adjustment

a. The NAVD 88 adjustment is the culmination of over ten years of work. This effort has included: establishing about 100,000 km of trunk line leveling to reinforce weak areas in the network; modernizing the vast amount of leveling observation and description data that has been collected for over a century; performing adjustments of sections of the network to verify the quality of the observation data; informing the public users of the network of the pending change and determining the impact on the nation's engineering activities. After the datum definition was selected to be a minimally constrained adjustment, the final task of this effort was to perform the least squares adjustment of the whole network.

b. The general adjustment of NAVD 88 was completed in June 1991. The primary network consists of 200 loops containing 909 junction BMs. The network connects to 57 primary tidal stations, which are part of the National Primary Tidal Network, and 55 international water-level stations along the Great Lakes. In addition, 28 border connections were made to the Canadian vertical control network and 13 to the Mexican vertical control network. Third Order BMs of other agencies (e.g., USACE) were not included in this adjustment. The 500,000 BMs established by the USGS were also not placed in computer readable form and therefore will not have NAVD 88 heights. In addition, USACE commands have established thousands of BMs that will not have NAVD 88 heights.

c. A particular concern for the developers of the NAVD 88 was how to resolve the many issues associated with the National Mapping Program (NMP) of USGS and the National Map Accuracy Standard (NMAS). The NMP includes more than 83,000 different map products of which over 7 million copies are distributed annually. Almost all of these products contain elevation information as contours and spot elevations on maps or as elevation arrays in Digital Elevation Models. Changing these products, both graphical and digital, to the NAVD 88 will be a massive and costly undertaking and will require a decade or more to complete.

d. The new leveling data have additional corrections applied for refraction and rod correction and are adjusted in geopotential units rather than the orthometric system used in the past. The datum definition is the most scientifically acceptable of all the definitions considered and is the most natural because it is based on an undisturbed representation of the Earth's gravity field. It is the most suitable for the geoid height computations needed for the reduction of GPS ellipsoidal heights. The main disadvantage is the differences with MSL on the west coast. At Seattle, a person standing on the zero elevation contour (NAVD 88) will barely have their head above water at mid-tide.

e. Preliminary analysis indicates that the overall differences between orthometric heights referred to NAVD 88 and to the NGVD 29 range from approximately -40 cm to +150 cm. Most surveying applications should not be significantly affected because the changes in relative height between adjacent BMs in most geotechnically stable areas should be less than 1 cm. In many geotechnically stable areas, a single bias factor describing the difference between NGVD 29 and NAVD 88 can be estimated and used for most mapping applications. This was a significant consideration for assessing the impact on the national mapping products. The absolute height values will change much more, but this should not be the surveyor's biggest concern, since he/she should be concerned with ensuring that all height values of BMs are referenced to the same vertical datum. The overall differences between dynamic heights referred to the IGLD 85 and to IGLD 55 will range from approximately 1 to 40 cm.

C-9. Maintenance of Parallel Datums by NGS: NAVD 88 and NGVD 29

For a period of time, NGS will support both the NAVD 88 and the NGVD 29. Continued maintenance of NGVD 29 will depend on user demands and budget constraints.

C-10. International Great Lakes Datum of 1955 (IGLD 55)

a. IGLD 55 is a datum common to the United States and Canada and is defined by international agreement. Before the establishment and adoption of the IGLD 55, the differences in elevation between the lakes had been determined but had not been connected to sea level and lake level data published from the United States and Canada did not match for the same lakes and rivers. The IGLD 55 was an international cooperative effort between the those two countries, the result of which was that the Great Lakes-St. Lawrence River system was then covered by a single uniform vertical control network. The IGLD 55 is different and separate from the NGVD 29.

b. IGLD 55 is by definition a hydraulic (i.e., dynamic) datum. The reference zero for IGLD 55 is based on mean water level surface at Father's Point (Pointe-au-Pere), Québec, Canada. Holding this point fixed determined the IGLD 55 datum. A procedure termed a "water level transfer" has been used to establish a local vertical datum on each of the Great Lakes. Research has concluded that the water level transfer technique was concluded to be at least as accurate as First Order, Class I geodetic leveling. The remaining lakes were incorporated using a combination of level lines and water level transfers. Adjusted elevations on the IGLD55 are referenced using the dynamic number system. The dynamic value of a benchmark (BM) is not a true linear elevation, but a serial number given to the level surface on which the mark lies. Dynamic elevations were adopted for the IGLD 55 primarily because they provide a means by which the geopotential hydraulic head can be measured more accurately between two points.

c. The earth's crust experiences movements around the entire Great Lakes and St. Lawrence River area. Therefore, the vertical reference datum for this area must be vertically readjusted every 25 to 30 years. This crustal movement is called "isostatic rebound," which is the gradual rising of the earth rebounding from the weight of the glaciers during the last glacial age. When IGLD 55 was created, it was known that readjustment would be necessary due to the effects of isostatic rebound. Crustal movement is not uniform across the Great Lakes basin and causes bench marks to shift not only with respect to each other, but also with respect to the initial reference point. Subsidence and other local effects can cause bench marks to shift as well.

C-11. International Great Lakes Datum of 1985 (IGLD 85)

a. The Coordinating Committee on Great Lakes Basic Hydraulic and Hydrologic Data revised the IGLD 55 datum and established IGLD 85. This committee has input to the international management of the Great Lakes-St. Lawrence River system. Representatives from the US and Canada are members on this committee. The efforts of the Coordinating Committee to revise IGLD 55 and establish IGLD 85 were coordinated with the efforts to establish the new common international vertical datum for the US, Canada and Mexico, NAVD 88.

b. The IGLD 85 is the current vertical control reference system in the Great Lakes Basin. The IGLD 55 was the vertical control reference system for this area until the publication of the IGLD 85 in January, 1992. Originally, an IGLD 80 had been planned, but the project was extended. The epoch that was actually used to determine the mean water level for the new datum was from 1983-1988 of which the mean year is 1985. The reference zero point for IGLD 85 is located at benchmark #1250G, located at Rimouski, Québec. This benchmark has an IGLD85 elevation of 6.723 meters and IGLD 55 elevation of 6.263 meters. IGLD 85 increases the number and accuracy of benchmarks in the Great Lakes area.

Corps districts were targeted to have converted over to IGLD 85 by January of 1993. This cannot be practically done until NOAA publishes complete IGLD 85 BM data for the area. IGLD 85 data is available for the gages, but the spacing of the gages is not dense enough to support conversion of local project control.

c. Agencies in the US and Canada will use IGLD 85. The National Oceanic Service (NOS) and the Canadian Hydrographic Service (CHS) began reporting water levels referenced to IGLD 85 upon its implementation in January, 1992. For a period of time, conversion factors for both IGLD 55 and IGLD 85 water level data will be provided by NOAA/NOS Great Lakes Section and CHS. The monthly water level bulletins published by USACE and CHS will reflect this information.

d. IGLD 85 will not change water levels established for federal flood insurance programs in the US. These levels will be referred to NAVD 88. Elevations common to both NAVD 88 and IGLD 85 are available from NOAA. Lake level outflows also will not be affected by the datum change to IGLD 85. As benchmark information becomes available, navigation, construction, and other improvement work on the Great Lakes should be referred to IGLD 85. Either datum is acceptable until the benchmark data is available for the respective USACE District or project area. Drawings shall include a note for the vertical IGLD datum in use to avoid blunders. USACE permit applications will still be referenced to the Ordinary High Water Mark (OHWM) as defined under Section 10 of the Rivers and Harbors Act. As benchmark information becomes available, new applications should reference IGLD 85.

C-12. NAVD 88 and IGLD 85

a. The specific needs of the Great Lakes system were taken into account while the decision were being made about how NAVD 88 was to be established. Analyses of data in the Great Lakes Basin was used to determine the effects of the datum constraint, magnitudes of height changes from the IGLD 55, deficiencies in the network design, selection of water-level station pairs to be used to generate zero geopotential difference observations, and additional re-leveling requirements. This coordination provided a check on the accuracy of the work and established a conversion between the IGLD 85 and NAVD 88.

b. Elevations referenced to NGVD 29 are unacceptable for use in resolving the involved problems of the Great Lakes System. The reference zero for NGVD 29 is not located within the Great Lakes system and orthometric elevations are not sufficient for use with large bodies of water such as the Great Lakes. The dynamics of large bodies of water are not modeled well by considering them as single equipotential surfaces. Other forces such as gravity must be considered. For example, water level measurements obtained at both ends of the Lake and connected to the NGVD 29 would show some magnitude of a permanently northerly slope.

c. The general adjustment of the NAVD 88 and IGLD 85 is one and the same. A minimum constraint adjustment of Canadian-Mexican-US leveling observations was performed holding fixed the height of the primary tidal BM, referred to the new IGLD 85 local mean sea level height value, at Father's Point/Rimouski, Québec, Canada. This constraint satisfies the requirements of shifting the datum vertically to minimize the impact of NAVD 88 on USGS mapping products, as well as provides the datum point desired by the IGLD Coordinating Committee for IGLD 85. The only difference between IGLD 85 and NAVD 88 is that IGLD 85 BM values are given in dynamic height units and NAVD 88 values are given in Helmert orthometric height units. The geopotential numbers of BMs are the same in both systems.

d. Geopotential numbers from the general adjustment of NAVD 88 were used to compute IGLD 85 dynamic heights. They will provide the best estimate of hydraulic head. If secondary gage data are placed in computer readable form, they will also be incorporated into NAVD 88 and IGLD 85. NGS will publish NAVD 88 heights and provide, upon special request, geopotential numbers for all BMs included in NAVD 88.

e. The use of GPS data and a high-resolution geoid model to estimate accurate GPS-derived orthometric heights will be a continuing part of the implementation of NAVD 88 and IGLD 85. It is important that users initiate a project to convert their products to NAVD 88 and IGLD 85. The conversion process is not a difficult task, but will require time and resources. Other local reference planes have been established by local jurisdictions and these can be referenced to either IGLD 85 or NAVD 88.

C-13. 1974 Low Water Reference Plane (LWRP)

a. On the Mississippi River between the mouths of the Missouri and the Ohio Rivers (the Middle Mississippi River), depths and improvements are referenced to a LWRP. No specific LWRP year is used for the Middle Mississippi north of Cairo, IL. Below Cairo, IL, depths and improvements along the Lower Mississippi River are referenced to the 1974 LWRP. This is also a hydraulic reference plane, established from long term observations of the river's stage, discharge rates, and flow duration periods. The low water profile was developed about the 97-percent flow duration line. The elevation of the 1974 LWRP drops gradually throughout the course of the Mississippi, however, some anomalies in the profile are present in places (particularly in areas containing rock bottoms or groins/weirs). The gradient is approximately 0.5 feet per river mile. The ever-changing river bottom will influence the 1974 LWRP. Changes in the stage-discharge relationship will influence the theoretical flow line for the 1974 LWRP.

b. Construction and improvements along the lower river are performed relative to the 1974 LWRP at a particular point. Differences in 1974 LWRP elevations between successive points along the river are determined from simultaneous staff readings and are referenced to benchmarks along the bank. The 1974 LWRP slope gradients between any two points must be corrected by linear interpolation of the profile. Thus, over a typical 1-mile section of river with a 0.5-foot gradient, each 1000-foot C/C river cross section will have a different 1974 LWRP correction, each dropping successively at approximately 0.1-foot increments.

c. Where practicable and feasible, NAVD 88 should be used as the common reference plane from which 1974 LWRP elevations are measured. The relationship of all project datums to both NGVD 29 and NAVD 88 should be clearly noted on all drawings, charts, maps, and elevation data files. All initial surveys should be referenced to both NAVD 88 and NGVD 29. If this is not feasible, then NGVD 29 should be used as the common reference plane from which 1974 LWRP elevations are measured until the move to NAVD 88 can be accomplished. Differences between the 1974 LWRP and NGVD 29 are published for the reference benchmarks used to control surveys and construction activities. In some districts, surveys are performed directly on NGVD 29 without regard to the 1974 LWRP profile (i.e., elevations above NGVD 29 are plotted rather than depths below 1974 LWRP). The 1974 LWRP depths are then contoured from the plotted NGVD 29 elevations, with the 1974 LWRP profile gradients applied during the contouring process. If a survey is conducted over a given reach of the river, the 1974 LWRP-NAVD 88 and/or the 1974 LWRP-NGVD 29 conversion must be interpolated based on the slope profile over that reach.

d. Controlled portions of the Upper Mississippi are referred to pool levels between the controlling structures. Although a variety of reference datums are used on other controlled river systems or impoundment reservoirs, most are hydraulically based and relate to some statistical pool level (e.g., "normal pool level," "flat pool level," "minimum regulated pool level ", etc.).

e. On the Mississippi River above Melvin Price Locks and Dam at Alton, IL, to Lock and Dam No. 22 at Saverton, MO, the reference used is related to the minimum regulated pool elevation. These pools are regulated referenced to a "hinge point". The pools are drawn down when the river's flow will provide adequate navigation depths naturally. When the flows are reduced to low volumes, the pools are reestablished and are essentially level. The depths and improvements along this reach of the Mississippi River are referenced to the "minimum regulated pool" elevations.

f. On the Mississippi River above Lock and Dam No. 22 at Saverton, MO, to St. Paul, MN, a "flat pool level" reference is used, and soundings are shown as "depth below flat pool". The flat pool is the authorized elevation of the navigation project and can be referenced to any number of local datums. Most commonly, this level is referenced to the mean sea level (MSL) datum of 1912, the general adjustment which preceded 1929. Conversions between MSL 1912 and NGVD 29 are available. The Illinois Waterway pool elevations are referred to NGVD 29, however, relationships to numerous other datums are also made.

g. Vertical clearances (bridges, transmission lines, etc.) are usually measured relative to high and low waters of record, or relative to full pool elevations. Shore lines shown on river drawings and navigation maps may be referenced to a low water datum (i.e., 1974 LWRP). On the Mississippi River above Lock and Dam No. 22 at Saverton, MO, the plotted shore line is referenced to full pool stage at dams with discharges equaled or exceeded 90 percent of the time. Given the variety of reference levels, special care must be taken to properly identify the nature and source of all vertical reference datums used on a project. The datum notes should include and clearly depict the relationship to NAVD 88.

C-14. NAVD 88 and the National Mapping Program (NMP)

a. The NMP of the US Geological Survey (USGS) includes more than 83,000 different map products. Some 55,000 of these are in the 7.5-minute, 1:24,000-scale, primary quadrangle map series (7.5-minute quads). These maps are widely used by Corps planners. Since the 7.5-minute quad series is the largest scale in the NMP and contains the greatest detail and elevation accuracy, it will be significantly affected by the vertical datum change.

b. The contour intervals of the 7.5-minute map series are selected to best express the topography of the area. With a few exceptions, the contour intervals range from 5 feet for very flat country to 80 feet for rugged mountainous terrain. In between these limits are 10-, 20-, and 40-foot contour intervals. A few maps in recent years have been compiled with metric value contours. The USGS production processes were designed to produce maps that meet the requirements of the National Map Accuracy Standards (NMAS). This standard requires that the error in 90% of the test points be less than one-half contour interval. Field survey methods are generally used to test the maps, and the elevation on the map is determined by interpolation between contours.

c. Other forms of vertical information shown on the USGS 7.5-minute maps are BMs and useful elevations, which are indicated by a cross symbol with the elevation given to the nearest foot. These elevations are established by geodetic leveling of Third Order accuracy or better. Spot elevations are measured by field or photogrammetric methods are readily identifiable features, eg: natural lakes, definite tops and saddles, fence corners, or road intersections. These elevations are usually placed at a density of about one-per-square mile and are considered to be accurate to within three-tenths of the contour interval.

d. Digital files of topographic information will also be affected by a change in the vertical datum. A Digital Elevation Model (DEM) consists of a sampled array of elevations for ground positions that are usually at regularly spaced intervals. For the 7.5-minute DEM, the horizontal framework is the Universal Transverse Mercator (UTM) system and the spacing is 30 meters. The 1-degree DEM horizontal coordinate system is based on the latitude and longitude positions of the World Geodetic System of 1972

(WGS 72) datum and the spacing is 3 arc-seconds. Another form of the digital topographic data that will be affected by the datum conversion is the hypsography category Digital Line Graphs (DLG).

e. In support of the production of the USGS topographic maps, a 3rd Order level network was established that resulted in few places being more than 5 miles from basic vertical control. These lines were usually established along farm roads, railroads, desert track roads, and mountain trails (less dense in mountain areas). USGS field surveyors have established nearly 500,000 BMs. Most of this work is on NGVD 29.

f. Changing the above NMP products, both graphic and digital, to the NAVD 88 will be a massive and costly undertaking and will require a decade or more to complete.

(1) In areas where the datum change is very small compared with the contour interval, advantage can be taken of the tolerance in the NMAS (i.e., 90% of the test point errors are less than ½ of the contour interval). If the datum change is only 1/10th of the contour interval, then the existing contours will still meet NMAS and will not require recompilation. The labeled elevations for BMs and spot elevations will need to be changed. This type of conversion is a low-cost approach but might be useful to extend the life of an otherwise sound map series. This is not a technically correct solution because a small bias is being introduced. Special care must be taken to insure that the contouring is in agreement with changed labeled elevations.

(2) Recompiling the contours and spot elevations on a 7.5-minute quad map is the most geometrically correct method of fitting the new vertical datum; however, this is an expensive approach. Therefore, total recompilation and recontouring due to an out-of-date datum is not considered to be cost-effective. However, some other factors may justify recompilation:

- Change to metric contours
- Terrain changes because of subsidence or other causes
- Inaccurate existing contours or inappropriate contour interval

g. Adjusting the USGS 3rd Order leveling network to the NAVD 88 is a different challenge, because high accuracy is needed to maintain its usefulness as geodetic data. This level of accuracy can only be provided by a least squares adjustment of the old observations to the new NAVD 88 primary network elevations.

h. The selected NAVD 88 datum definition best fits the needs of the NMP. The important characteristics are:

- Small elevation changes for the eastern half of the USA where the 7.5 map contour intervals are the smallest and large changes for the western half where contour intervals are the largest.

- The isograms representing these changes are smoother and show less irregularities.

Both of these are important if the map patching conversion techniques is to be used. The requirements are that the shift values be small compared with the contour interval and the gradient in the datum change be minimal so that a single change value can be applied over an entire 7.5-minute map with little noticeable error. A vertical shift (bias) in the defining elevation is desirable to expand the favorable interval/elevation change area over the entire US.

C-15. National Flood Insurance Program (NFIP) Transition to NAVD 88

a. The NFIP is a federal program that provides identification of flood hazard areas on a community basis and includes availability of insurance against flood damages. When a community joins the NFIP, it agrees to adopt minimum Federal floodplain management criteria enforced by local regulations. The major product of the NFIP is the Flood Insurance Study (FIS) and accompanying Flood Insurance Rate Map (FIRM), which describes a community's floodplains and regulatory floodways, as well as computed flood profiles. FEMA prepares an FIS for severely flood prone communities, that identifies 100-year base flood elevations (BFEs) and flood hazard areas. Nearly all flood maps for these areas are referenced to the NGVD 29. Conversion to NAVD 88 will require the education of map users as well as map producers. New FISs will be based on NAVD 88. Existing studies and maps will be converted when substantive revisions occur to redefine flood hazards.

b. The dual mission of the NFIP through the Federal Insurance Administration (FIA) is to reduce future flood losses as well as provide insurance coverage to offset deficit producing disaster assistance payments. To provide the floodplain management information to the communities, FEMA provides flood hazard mapping of major flooding sources within a community, usually based on the 100 year BFEs.

c. FEMA will be converting its products the NAVD 88 in a gradual process primarily as FEMA's FISs and FIRMs are republished. Currently, the vast majority of FEMA products for the NFIP are referred to NGVD 29. All FEMA studies contracted for FY93 and beyond will require the use of NAVD 88 as vertical control. Since October 1992, all requests for map change actions received required the inclusion of NAVD 88 data. The NFIP will transition to NAVD 88 on a project basis or as other reasons for revision indicate.

C-16. Effects of NAVD 88 on FIRMs and Communities

a. The base 100-year BFEs and Elevation Reference Marks (ERMs) will be converted to NAVD 88. Use of the current datum will be acceptable until the change is made of the FIRM. After that time, all flood insurance policy sales and renewals will be based on elevations referenced to NAVD 88. Determination of locations of structures and proposed projects with respect to special flood hazard areas (SFHAs) will be based upon elevations referenced to NAVD 88. The datum listed as the reference datum on the applicable FIRM panel should be used for Elevation Certificate completion. This is FEMA policy for all NFIP communities since the NAVD 88 is defined over all these regions. For Hawaii, the Pacific Trust Territories, the Commonwealth of Puerto Rico and the US Virgin Islands, their datums will be adjusted based upon re-leveling work done in those areas and their 1960-78 tidal epoch. Their local datums will be designated as NAVD 88 and will be defined by their current local MSL determinations.

b. The conversion of the vertical reference datum from NGVD 29 to NAVD 88 will take place over time and documentation should be carefully maintained to reflect which vertical reference datum was used. If a FIS is completed with ERMs referenced to NAVD 88, the conversion method and results shall be part of the deliverable items with that FIS. The FEMA Project Officer should be consulted to decide if NAVD 88 must be used or not, prior to commencement of any work for that FIS.

c. No requirement is made by FEMA on local communities to relevel their local vertical control networks for NFIP purposes. If, however, the community's current vertical control system is referenced to an NGS bench mark(s) that is included in the new data available from NGS, then appropriate conversion can be made. Questions regarding the mechanics of shifting to NAVD 88 may be addressed to the appropriate FEMA Regional Offices.

d. A very important potential problem is caused by mixing datums. If a consistent datum is used for determining Base 100-year BFEs and lowest floor elevation, actuarial rating and building requirements would be correctly determined. If mixed datums are used, significant problems arise. For example, if the 1-foot NAVD 88 BFE is used mixed with the 2-foot NGVD 29 lowest floor elevation, an error of 2 feet (not in the NFIP's favor) would occur. There are roughly 22 million people living in the nation's floodplains. Proposed elevations above the BFE are sometimes minimal for economic considerations, so up-to-date and properly referenced elevation data are a must. Correct referencing of those floodplains to the nation's vertical datum must be accomplished as soon as it is administratively and financially possible to do so.

C-17. Map Revision Requests to FEMA

a. As the NFIP moves further into the maintenance stage of the mapping program, the major action with respect to mapping of flood hazards will be the review and processing of requests for revisions to the currently effective FISs and FIRMs. The decision of whether to convert to NAVD 88 as the reference datum for a revision action must be made on a case by case basis. All map revision requests should contain documentation of vertical control used and those requests to FEMA after 1 October 1992 must include vertical control data referenced to NAVD 88. The decision regarding the published reference for the map revision will be made by the FEMA Project Officer for the applicable region.

b. Map revision requests shall include either NGS BM data or the method and computations used to tie to NGS BMs. If NGS BM data is unavailable, documentation to that effect must be submitted with the map revision request. If a computer program was used, the name of the program should be included along with where an exact copy of that program can be obtained. The leveling field notes should also be included. All surveying data must be certified by a licensed Land Surveyor or registered Professional Engineer. For all map revisions, the datum referenced on the current FIRM shall be used unless otherwise directed by the Project Officer. When the current map datum is used, a conversion factor to allow comparison to NAVD 88 elevation should be included.

C-18. Impact of NAVD 88 Change on Flooding Sources

a. Riverine and lacustrine flooding. For most areas affected by this type of flooding, the changes from NGVD 29 to NAVD 88 will be adequately addressed by a shift factor for areas of USGS 7.5-minute quadrangle series topographic maps. In larger areas than that, such as county wide studies involving significant stream or river reaches, additional considerations are necessary.

b. Coastal flooding. In areas affected by coastal flooding, additional care must be taken to avoid confusion with local MSL data that are used in addition to NGVD 29 data. In many areas, these are assumed to be the same, while in other areas differences may exist based upon the latest tidal observations. NGS has included in its published data the new reference elevations for tidal stations previously taken as 0.0000 foot on NGVD 29. All vertical data must clearly reflect what basis of vertical control was used. For example, if mean low water datum was used, conversion to NAVD 88, or at a minimum NGVD 29, must be provided and sealed by a certified land surveyor.

c. Other influences. Areas that have experienced crustal motion or land subsidence since the publication of the vertical control data for that area, must be referenced to at least one BM known to be stable. Documentation from a certified land surveyor or by the agency that recently releveled the BM must be included with any data submitted to FEMA. NGS will be publishing special reports for these areas as part of its ongoing long term task.

(1) For areas that experience uniform change over a given range, and where datum difference can be expressed as a bias factor for specific geographic areas, little if any distortion will occur in the hydrologic and hydraulic parameters that influence the definition of a given floodplain. In these instances, FEMA will be concerned with assuring that the proper conversion of ground and hydraulic elevations takes place.

(2) In instances where nonuniform elevation differences are indicated, an investigation of the potential effect on hydraulic behavior will be required. Usually, unless the change in flood elevation or depth is greater than 0.5 foot or in some cases 1.0 foot, no republication of the flood elevations is dictated. Indications of potential changes of 1.0 foot or more will probably place the stream or community on the priority list for a contracted restudy to establish the exact effect of the changes.

C-19. FEMA Policy for Map Conversion: Affect on NFIP Products

FEMA datum conversion activities called for all FY93 FIS's to be referenced to NAVD 88 and that this action be confirmed by the contractor with the Project Officer prior to the beginning of survey work. The study contractor is responsible for assuring that proper reference to NAVD 88 is made.

C-20. NAVD 88 Requirements for Flood Insurance Studies

a. Type 15 studies. For initial studies, NAVD 88 shall be required. Exceptions must be approved by the Project Officer prior to the start of survey work.

b. Type 19 studies and limited map maintenance program studies. Use of NAVD 88 shall be the decision of the Project Officer. If NGVD 29 is used, then a conversion factor to NAVD 88 should be included with the study material.

c. The use of NAVD 88 for these studies will be determined by the extent of the changes that will occur to the community's FIRM when revised. For communities whose FIRM is larger than 1 panel and revision of other panels is unlikely with the restudy, the use of NGVD 29 may be continued, but a note explaining the conversion to NAVD 88 should be included in the "NOTES" in the map border of the panel being revised.

Appendix D
Requirements and Procedures for Referencing Coastal Navigation Projects to Mean Lower Low Water (MLLW) Datum

D-1. Purpose

This appendix provides guidance, technical considerations, and general implementation procedures for referencing coastal navigation projects to a consistent Mean Lower Low Water (MLLW) datum based on tidal characteristics defined and published by the US Department of Commerce. This guidance is necessary to implement applicable portions of Section 224 of the Water Resources Development Act of 1992.

D-2. Applicability

This technical guidance in this appendix applies to Commands having responsibilities for design of river and harbor navigation projects on the Atlantic, Gulf, and Pacific coasts, and where such projects are subject to tidal influence.

D-3. References

   a. Rivers and Harbors Appropriation Act of 1915 (38 Stat. 1053; 33 U.S.C. 562).

   b. Water Resources Development Act of 1992 (WRDA 92), Section 224, Channel Depths and Dimensions.

   c. EM 1110-2-1003, Hydrographic Surveying.

   d. EM 1110-2-1414, Water Levels and Wave Heights for Coastal Engineering and Design.

   e. Tidal Datum Planes, Special Publication 135, US Department of Commerce.

   f. Manual of Tide Observations, Publication 30-1, US Department of Commerce.

   g. Tide and Current Glossary, US Department of Commerce.

   h. Statement of Work for the Installation, Operation and Maintenance of Tide Stations, US Department of Commerce.

   i. User's Guide for the Installation of Bench Marks and Leveling Requirements for Water Level Stations, US Department of Commerce.

   j. The National Tidal Datum Convention of 1980, US Department of Commerce.

D-4. Background

a. Depths of USACE navigation projects in coastal areas subject to tidal influences are currently referred to a variety of vertical reference planes, or datums. Most project depths are referenced to a local or regional datum based on tidal phase criteria, such as Mean Low Water, Mean Lower Low Water, Mean Low Gulf, Gulf Coast Low Water Datum, etc. Some of these tidal reference planes were originally derived from US Department of Commerce, National Ocean Service (NOS) observations and definitions used for the various coasts. Others were specifically developed for a local project and may be without reference to an established vertical network (e.g., National Geodetic Vertical Datum of 1929) or a tidal reference. Depending on the year of project authorization, tidal epoch, procedures, and the agency that established or connected to the reference datum, the current adequacy of the vertical reference may be uncertain, or in some cases, unknown. In some instances, project tidal reference grades may not have been updated since original construction. In addition, long-term physical effects may have significantly impacted presumed relationships to the NOS MLLW datum.

b. The National Tidal Datum Convention of 1980 established one uniform, continuous tidal datum for all marine waters of the United States, its territories, and Puerto Rico. This convention thereby lowered the reference plane (and tidal definition) of both the Atlantic and Gulf coasts from a mean low water datum to a MLLW datum. In addition, the National Tidal Datum Epoch was updated to the 1960-1978 period and mean higher/high water datums used for legal shoreline delineation were redefined.

c. Since 1989, nautical charts published by the US Coast and Geodetic Survey (USC&GS), US Department of Commerce, reference depths (or soundings) to the local MLLW reference datum, also termed a "chart datum." US Coast Guard (USCG) Notices to Mariners also refer depths or clearances over obstructions to MLLW. Depths and clearances reported on USACE project/channel condition surveys provided to USC&GS, for incorporation into their published charts in plan or tabular format, must be on the same NOS MLLW reference as the local chart of the project site.

d. WRDA 92, Section 224, requires consistency between USACE project datums and USC&GS marine charting datums. This act amended Section 5 of the Rivers and Harbors Appropriation Act of 1915 to define project depths of operational projects as being measured relative to a MLLW reference datum for all coastal regions. (Only the Pacific coast was previously referenced to MLLW). The amendment states that this reference datum shall be as defined by the Department of Commerce for nautical charts (USC&GS) and tidal prediction tables (NOS) for a given area. This provision requires USACE project reference grades be consistent with NOS MLLW.

D-5. Impact of MLLW Definition on USACE Projects

a. Corps navigation projects that are referenced to older datums (e.g., Mean Low Water along the Atlantic coast or various Gulf coast low water reference planes) must be converted to and correlated with the local MLLW tidal reference established by the NOS. Changes in project grades due to redefining the datum from mean low water to NOS MLLW will normally be small, and in many cases will be compensated for by offsetting secular sea level or epochal increases occurring over the years. Thus, impacts on dredging due to the redefinition of the datum reference are expected to be small and offsetting in most cases.

b. All Corps project reference datums, including those currently believed to be on MLLW, must be checked to insure that they are properly referred to the latest tidal epoch, and that variations in secular sea level, local reference gage or benchmark subsidence/uplift, and other long-term physical phenomena are properly accounted for. In addition, projects should be reviewed to insure that tidal phase and range characteristics are properly modeled and corrected during dredging, surveying, and other marine construction activity, and that specified project clearances above grade properly compensate for any tidal range variances. Depending on the age and technical adequacy of the existing MLLW reference (relative to NOS MLLW), significant differences could be encountered. Such differences may dictate changes in channels currently maintained. Future NOS tidal epoch revisions will also change the project reference planes.

c. Conversion of project datum reference to NOS MLLW may or may not involve field tidal observations. In many projects, existing NOS tidal records can be used to perform the conversion, and short-term simultaneous tidal comparisons will not be required. Tidal observations and/or comparisons will be necessary for projects in areas not monitored by NOS or in cases where no recent or reliable observations are available.

D-6. Implementation Actions

A number of options are available to USACE commands in assessing individual projects for consistency and accuracy of reference datums, and performing the necessary tidal observations and/or computations required to adequately define NOS MLLW project reference grade. Datum establishment or verification may be done using USACE technical personnel, through an outside Architect-Engineer contract, by another Corps district or laboratory having special expertise in tidal work, or through reimbursable agreement with NOS. Regardless of who performs the tidal study, all work should be closely coordinated with both the USC&GS and NOS in the Department of Commerce.

a. Technical specifications. The general techniques for evaluating, establishing, and/or transferring a tidal reference plane are fully described in the USACE and Department of Commerce publications referenced in paragraph D-3. These references should be cited in technical specifications used for a tidal study contract or reimbursable agreement with another agency/command.

b. Department of Commerce contacts. Before and during the course of any tidal study, close coordination is required with NOS. The NOS point of contact is the Chief, Ocean and Lake Levels Division, National Ocean Service (ATTN: N/OES2), 6001 Executive Blvd., Rockville, MD 20852, telephone (301) 443-8807, FAX (301) 443-1920.

c. Sources. If in-house forces are not used, the following outside sources may be utilized to perform a tidal study of a project, including any field tidal observations.

(1) Architect-Engineer (A-E) Contract. A number of private firms possess capabilities to perform this work. Either a fixed-scope contract or indefinite delivery type (IDT) contract form may be utilized. In some instances, this type of work may be within the scope of existing IDT contracts. Contact NOS to obtain a typical technical specification which may be used in developing a scope of work. The references in paragraph D-3 of this appendix must be cited in the technical scope of work for the A-E contract.

(2) Reimbursable Support Agreement. Tidal studies and datum determinations may be obtained directly from the NOS, Department of Commerce, via a reimbursable support agreement. A cooperative agreement can be configured to include any number of projects within a district. Funds are provided to NOS by standard inter-agency transfer methods and may be broken down to individual projects. Contact the Chief of the Ocean and Lake Levels Division at the address given in paragraph D-6b to coordinate and schedule a study agreement.

d. Scheduling of conversions. Section 224 of WRDA 92 did not specify an implementation schedule for converting existing projects to NOS MLLW (or verifying the adequacy of an existing

MLLW datum). It is recommended that a tidal datum study be initiated during a project's next major maintenance cycle.

   e. Funding. No centralized account has been established to cover the cost of converting projects to NOS MLLW datum. Project Operations and Maintenance funds will be used to cover the cost of tidal studies and/or conversions on existing projects. For new construction, adequate funding should be programmed during the initial planning and study phases. Budget estimates for performing the work can be obtained from NOS.

   f. MLLW relationship to national vertical network. USACE tidal benchmarks should be connected to the national vertical network maintained by USC&GS; either the National Geodetic Vertical Datum, 1929 Adjustment (NGVD 29) or the updated North American Vertical Datum, 1988 Adjustment (NAVD 88). Project condition surveys, maps, reports, studies, etc. shall clearly depict the local relationship between MLLW datum and the national vertical network.

   g. Changes in dredging. It is not expected that the datum conversion will significantly impact dredging requirements. USACE commands should request HQUSACE guidance should a datum conversion cause a significant change in a channel's maintained depth.

Glossary

1. Abbreviations

2D	Two-dimensional
2DRMS	Twice the distance root mean square
3D	Three-dimensional
ACSM	American Congress on Surveying and Mapping
BFE	Base Flood Elevation
BS	Backsight
CONUS	CONtinental United States
CORPSCON	CORPS CONvert
DEM	Digital Elevation Model
DLG	Digital Line Graph
DoD	Department of Defense
DGPS	Differential Global Positioning System
EDM	Electronic Distance Measurement
EFARS	Engineer Federal Acquisition Regulation Supplement
EM	Engineer Manual
E&D	Engineering and Design
FEMA	Federal Emergency Management Agency
FGCS	Federal Geodetic Control Subcommittee
FGDC	Federal Geographic Data Committee
FIRM	Flood Insurance Rate Map
FIS	Flood Insurance Study
FOA	Field Operating Activity
FS	Foresight
GIS	Geographic Information System
GPS	Global Positioning System
GRS 80	Geodetic Reference System of 1980
HARN	High Accuracy Regional Networks
HI	Height of Instrument
HQUSACE	Headquarters, US Army Corps of Engineers
IDT	Indefinite Delivery Type
IGLD 55	International Great Lakes Datum of 1955
IGLD 85	International Great Lakes Datum of 1985
MACOM	Major Army Command
MLLW	Mean Lower Low Water
MSL	Mean Sea Level
MSL1912	Mean Sea Level Datum of 1912
NAD 27	North American Datum of 1927
NAD 83	North American Datum of 1983
NADCON	North American Datum Conversion
NATO	North Atlantic Treaty Organization
NAVD 88	North American Vertical Datum 1988
NFIP	National Flood Insurance Program
NGRS	National Geodetic Reference System
NGS	National Geodetic Survey
NGVD 29	National Geodetic Vertical Datum 1929
NMAS	National Map Accuracy Standard
NMP	National Mapping Program

NOAA ....National Oceanic and Atmospheric Administration
NOS ....National Ocean Service
NVCN........................National Vertical Control Network
OCONUS ....Outside the Continental United States
PICES ....Periodic Inspection and Continuing Evaluation (of Completed CW) Structures
SI ....International System of Units
SPCS ....State Plane Coordinate System
TBM ....Temporary Benchmark
TM.............................Transverse Mercator
USC&GS ....US Coast & Geodetic Survey
US..............................United States
USACE ....US Army Corps of Engineers
UTM ....Universal Transverse Mercator
VERTCON ....VERTical CONversion
WGS 84 ....World Geodetic System of 1984

2. Terms

Absolute GPS
Operation with a single receiver for a desired position. This receiver may be positioned to be stationary over a point. This mode of positioning is the most common military and civil application.

Accuracy
The degree to which an estimated (mean) value is compatible with an expected value. Accuracy implies the estimated value is unbiased.

Adjustment
Adjustment is the process of estimation and minimization of deviations between measurements and a mathematical model.

Altimeter
An instrument that measures elevation differences usually based on atmospheric pressure measurements.

Altitude
The vertical angle between the horizontal plane of the observer and a directional line to the object.
Angle of Depression
A negative altitude.

Angle of Elevation
A positive altitude.

Angular Misclosure
Difference in the actual and theoretical sum of a series of angles.

Archiving
Storing of documents and information.

Astronomical Latitude
Angle between the plumb line and the plane of celestial equator. Also defined as the angle between the plane of the horizon and the axis of rotation of the earth. Astronomical latitude applies only to positions on the earth and is reckoned from the astronomic equator, north and south through 90E. Astronomical latitude is the latitude that results directly from observations of celestial bodies, uncorrected for deflection of the vertical.

Astronomical Longitude
Arbitrarily chosen angle between the plane of the celestial meridian and the plane of an initial meridian. Astronomical longitude is the longitude that results directly from observations on celestial bodies, uncorrected for deflection of the vertical.

Astronomical Triangle
A spherical triangle formed by arcs of great circles connecting the celestial pole, the zenith and a celestial body. The angles of the astronomical triangles are: at the pole, the hour angle; at the celestial body, the parallactic angle; at the zenith, the azimuth angle. The sides are: pole to zenith, the co-latitude; zenith to celestial body, the zenith distance; and celestial body to pole, the polar distance.

Atmospheric Refraction
Refraction of electromagnetic radiation through the atmosphere causing the line-of-sight to deviate from a straight path. Mainly temperature and pressure conditions determine the magnitude and direction of curvature affecting the path of light from a source. Refraction causes the ray to follow a curved path normal the surface gradient.

Azimuth
The horizontal direction of a line clockwise from a reference plane, usually the meridian. Often called forward azimuth to differentiate from back azimuth.

Azimuth Angle
The angle less than 180° between the plane of the celestial meridian and the vertical plane with the observed object, reckoned from the direction of the elevated pole. In astronomic work, the azimuth angle is the spherical angle at the zenith in the astronomical triangle, which is composed of the pole, the zenith and the star. In geodetic work, it is the horizontal angle between the celestial pole and the observed terrestrial object.

Azimuth Closure
Difference in arc-seconds of the measured or adjusted azimuth value with the true or published azimuth value.

Backsight
A sight on a previously established traverse or triangulation station and not the closing sight on the traverse. A reading on a rod held on a point whose elevation has been previously determined.

Barometric Leveling
Determining differences of elevation from measured differences of atmospheric pressure observed with a barometer. If the elevation of one station above a datum is known, the approximate elevations of other station can be determined by barometric leveling. Barometric leveling is widely used in reconnaissance and exploratory surveys.

Baseline
Resultant three-dimensional vector between any two stations with respect to a given coordinate system. The primary reference line in a construction system.

Base net
The primary baseline used for densification of survey stations to form a network.

Base Points
The beginning points for a traverse that will be used in triangulation or trilateration.

Base Control
The horizontal and vertical control points and coordinates used to establish a base network. Base control is determined by field surveys and permanently marked or monumented for further surveys.

Bearing
The direction of a line with respect to the meridian described by degrees, minutes, and seconds within a quadrant of the circle. Bearings are measured clockwise or counterclockwise from north or south, depending on the quadrant.

Bench mark
A permanent material object, natural or artificial, on a marked point of known elevation.

Best Fit
To represent a given set of points by a smooth function, curve, or surface which minimizes the deviations of the fit.

Bipod
A two-legged support structure for an instrument or survey signal at a height convenient for the observer.

Bluebook
Another term for the "FGCS Input Formats and Specifications of the National Geodetic Data Base".

Blunder
A mistake or gross error.

Bureau International de l'Heure
The Bureau was founded in 1919 and its offices since then have been at the Paris Observatory. By an action of the International Astronomical Union, the BIH ceased to exist on 1 January 1988 and a new organization, the International Earth Rotation Service (IERS) was formed to deal with determination of the Earth's rotation.

Cadastral Survey
Relates to land boundaries and subdivisions, and creates units suitable for transfer or to define the limitations of title. The term cadastral survey is now used to designate the surveys of the public lands of the US, including retracement surveys for identification and resurveys for the restoration of property lines; the term can also be applied properly to corresponding surveys outside the public lands, although such surveys are usually termed land surveys through preference.

Calibration
Determining the systematic errors in an instrument by comparing measurements with correct values. The correct value is established either by definition or by measurement with a device that has itself been calibrated or of much higher precision.

Cartesian Coordinates
A system with its origin at the center of the earth and the x and y and z axes in the plane of the equator. Typically, the x-axis passes through the meridian of Greenwich, and the z-axis coincides with the earth's axis of rotation. The three axes are mutually orthogonal and form a right-handed system.

Cartesian System
A coordinate system consisting of axes intersecting at a common point (origin). The coordinate of a point is the orthogonal distance between that point and the hyperplane determined by all axes. A Cartesian coordinate system has all the axes intersecting at right angles, and the system is called a rectangular.

Celestial Equator
A great circle on the celestial sphere with equidistant points from the celestial poles. The plane of the earth's equator, if extended, would coincide with that of the celestial equator.

Celestial pole
A reference point at the point of intersection of an indefinite extension of the earth's axis of rotation and the apparent celestial sphere.

Celestial sphere
An imaginary sphere of infinite radius with the earth as a center. It rotates from east to west on a prolongation of the earth's axis.

Central Meridian
A line of constant longitude at the center of a graticule. The central meridian is used as a base for constructing the other lines of the graticule. The meridian is used as the y-axis in computing tables for a State Plane Coordinate system. That line, on a graticule, which represents a meridian and which is an axis of symmetry.

Chain
Equal to 66 feet or 100 links. The unit of length prescribed by law for the survey of the US public lands. One acre equals 10 square chains.

Chained Traverse
Observations and measurements performed with tape.

Chaining
Measuring distances on the ground with a graduated tape or with a chain.

Chart Datum
Reference surface for soundings on a nautical chart. It is usually taken to correspond to a low water elevation, and its depression below mean sea level is represented by the symbol Z_o. Since 1989, chart datum has been implemented to mean lower low water for all marine waters of the US its territories, Commonwealth of Puerto Rico and Trust Territory of the Pacific Islands.

Chi-square Testing
Non-parametric statistical test used to classify the shape of the distribution of the data.

Chronometer
A portable timekeeper with compensated balance, capable of showing time with extreme precision and accuracy.

Circle Position
A prescribed setting (reading) of the horizontal circle of a direction theodolite, to be used for the observation on the initial station of a series of stations that are to be observed.

Circuit Closure
Difference between measured or adjusted value and the true or published value.

Clarke 1866 Ellipsoid
The reference ellipsoid used for the NAD 27 horizontal datum. It is a non-geocentric ellipsoid formerly used for mapping in North America.

Closed Traverse
Starts and ends at the same point or at stations whose positions have been determined by other surveys.

Collimation
A physical alignment of a survey target or antenna over a mark or to a reference line.

Collimation Error
The angle between the actual line of sight through an optical instrument and an alignment.

Compass Rule
The correction applied to the departure (or latitude) of any course in a traverse has the same ratio to the total misclosure in departure (or latitude) as the length of the course has to the total length of the traverse.

Confidence Level

Statistical probability (in percent) based on the standard deviation or standard error associated with the normal probability density function. The confidence level is assigned according to an expansion factor multiplied by the magnitude of one standard error. The expansion factor is based on values found in probability tables at a chosen level of significance.

Conformal
Map projection that preserves shape.

Contour
An imaginary line on the ground with all points at the same elevation above or below a specified reference surface.

Control
Data used in geodesy and cartography to determine the positions and elevations of points on the earth's surface or on a cartographic representation of that surface. A collective term for a system of marks or objects on the earth or on a map or a photograph whose positions or elevation are determined.

Control Densification
Addition of control throughout a region or network.

Control Monuments
Existing local control or benchmarks that may consist of any Federal, state, local or private agency points.

Control Point
A point with assigned coordinates is sometimes used as a synonym for control station. However, a control point need not be realized by a marker on the ground.
Control Survey
A survey which provides coordinates (horizontal or vertical) of points to which supplementary surveys are adjusted.

Control Traverse
A survey traverse made to establish control.

Conventional Terrestrial Pole (CTP)
The origin of the WGS 84 Cartesian system is the earth's center of mass. The Z-axis is parallel to the direction of the CTP for polar motion, as defined by the Bureau of International de l'Heure (BIH), and equal to the rotation axis of the WGS 84 ellipsoid. The X-axis is the intersection of the WGS 84 reference meridian plane and the CTP's equator, the reference meridian being parallel to the zero meridian defined by the BIH and equal to the X-axis of the WGS 84 ellipsoid. The Y-axis completes a right-handed, earth-centered, earth-fixed orthogonal coordinate system, measured in the plane of the CTP equator 90 degrees east of the X-axis and equal to the Y-axis of the WGS 84 ellipsoid.

Coordinate Transformation
A mathematical process for obtaining a modified set of coordinates through some combination of rotation of coordinate axes at their point of origin, change of scale along coordinate axes, or translation through space

CORPSCON
(Corps Convert) Software package (based on NADCON) capable of performing coordinate transformations between NAD 83 and NAD 27 datums.

Crandall Method
Traverse misclosure in azimuth or angle is first distributed in equal portions to all the measured angles. The adjusted angles are then held fixed and all remaining coordinate corrections distributed among the distance measurements.

Cross sections
A survey line run perpendicular to the alignment of a project, channel or structure.

Curvature
The rate at which a curve deviates from a straight line. The parametric vector described by dt/ds, where t is the vector tangent to a curve and s is the distance along that curve.

Datum
Any numerical or geometrical quantity or set of such quantities which serve as a reference or base for other quantities.

Declination
The angle, at the center of the celestial sphere, between the plane of the celestial equator and a line from the center to the point of interest (on a celestial body).

Deflection of the Vertical
The spatial angular difference between the upward direction of a plumb line and the normal to the reference ellipsoid. Often expressed in two orthogonal components in the meridian and the prime vertical directions.

Deflection Traverse
Direction of each course measured as an angle from the direction of the preceding course.

Deformation Monitoring
Observing the movement and condition of structures by describing and modeling its change in shape.

Departure
The orthogonal projection of a line onto an east-west axis of reference. The departure of a line is the difference of the meridional distances or longitudes of the ends of the line.

Differential GPS
Process of measuring the differences in coordinates between two receiver points, each of which is simultaneously observing/measuring satellite code ranges and/or carrier phases from the NAVSTAR GPS constellation. Relative positioning with GPS can be performed by a static or kinematic modes.

Differential Leveling
The process of measuring the difference of elevation between any two points by spirit leveling.

Direction
The angle between a line or plane and an arbitrarily chosen reference line or plane. At a triangulation station, observed horizontal angles are referred to a common reference line and termed horizontal direction. A line, real or imaginary, pointing away from some specified point or locality toward another point. Direction has two meanings: that of a numerical value and that of a pointing line.

Direct Leveling
The determination of differences of elevation through a continuous series of short horizontal lines. Vertical distances from these lines to adjacent ground marks are determined by direct observations on graduated rods with a leveling instrument equipped with a spirit level.

Distance Angle
An angle in a triangle opposite a side used as a base in the solution of the triangle, or a side whose length is to be computed.

Dumpy Level
The telescope permanently attached to the leveling base, either rigidly to by a hinge that can be manipulated by a micrometer screw.

Earth-Centered Ellipsoid
Center at the Earth's center of mass and minor semi-axis coincident with the Earth's axis of rotation.

Easting
The distance eastward (positive) or westward (negative) of a point from a particular meridian taken as reference.

Eccentricity
The ratio of the distance from the center of an ellipse to its focus on the major semi-axis.

Electronic Distance Measurement (EDM)
Timing or phase comparison of electro-magnetic signal to determine an interferometric distance.

Elevation
The height of an object above some reference datum.

Ellipsoid
Formed by revolving an ellipse about its minor semi-axis. The most commonly used reference ellipsoids in North America are: Clarke 1866, Geodetic Reference System of 1980 (GRS 80), World Geodetic System of 1972 (WGS 72) and World Geodetic System of 1984 (WGS 84).

Ellipsoid height
The magnitude h of a point above or below the reference ellipsoid measured along the normal to the ellipsoid surface.

Error
The difference between the measured value of a quantity and the theoretical or defined value of that quantity.

Error Ellipse
An elliptically shaped region with dimensions corresponding to a certain probability at a given confidence level.

Error of Closure
Difference in the measured and predicted value of the circuit along the perimeter of a geometric figure.

Finite Element Method
Obtaining an approximate solution to a problem for which the governing differential equations and boundary conditions are known. The method divides the region of interest into numerous, interconnected sub-regions (finite elements) over which simple, approximating functions are used to represent the unknown quantities.

Fixed Elevation
Adopted as a result of tide observations or previous adjustment of spirit leveling, and which is held at its accepted value in any subsequent adjustment.

Foresight
An observation to the next instrument station. The reading on a rod that is held at a point whose elevation is to be determined.

Frequency
The number of complete cycles per second existing in any form of wave motion.
Geodesic Line
Shortest distance between any two points on any mathematically defined surface.

Geodesy
Determination of the time-varying size and figure of the earth by such direct measurements as triangulation, leveling and gravimetric observations.

Geodetic Control
Established and adjusted horizontal and/or vertical control in which the shape and size of the earth have been considered in position computations.

Geodetic Coordinates
Angular latitudinal and longitudinal coordinates defined with respect to a reference ellipsoid.

Geodetic Height
See Ellipsoid height.

Geodetic Latitude
The angle which the normal at a point on the reference ellipsoid makes with the plane of the equator.

Geodetic Leveling
The observation of the differences in elevation by means of a continuous series of short horizontal lines of sight.

Geodetic Longitude
The angle subtended at the pole between the plane of the geodetic meridian and the plane of a reference meridian (Greenwich).

Geodetic North
Direction tangent to a meridian pointing toward the pole defining astronomic north, also called true north.

Geodetic Reference System of 1980
Reference ellipsoid used to establish the NAD 83 system of geodetic coordinates.

Geoid
An equipotential surface of the gravity field approximating the earth's surface and corresponding with mean sea level in the oceans and its extension through the continents.

GPS (Global Positioning System)
DoD satellite constellation providing range, time, and position information through a GPS receiver system.

Gravimeter
Instrument for measuring changes in gravity between two points.

Gravity
Combined acceleration potential of an object due to gravitation and centrifugal forces.

Greenwich Meridian
The astronomic meridian through the center of the Airy transit instrument of the Greenwich Observatory, Greenwich, England. By international agreement in 1884, the Greenwich meridian was adopted as the meridian from which all longitudes, worldwide, would be calculated.

Grid Azimuth
The angle in the plane of projection between a straight line and the line (y-axis) in a plane rectangular coordinate system representing the central meridian. While essentially a map-related quantity, a grid azimuth may, by mathematical processes, be transformed into a survey- related or ground-related quantity.

Grid Inverse
The computation of length and azimuth from coordinates on a grid.

Grid Meridian
Line parallel to the line representing the central meridian or y-axis of a grid on a map. The map line parallel to the line representing the y-axis or central meridian in a rectangular coordinate system.

Gunter's Chain
A measuring device once used in land surveying. It was composed of 100 metallic links fastened together with rings. The total length of the chain is 66 feet. Also called a four-pole chain.

Gyrotheodolite
A gyroscopic device used to measure azimuth that is built-in or attached to a theodolite.

Histogram
A graphical representation of relative frequency of an outcome partitioned by class interval. The frequency of occurrence is indicated by the height of a rectangle whose base is proportional to the class interval.

Horizontal Control
Determines horizontal positions with respect to parallels and meridians or to other lines of reference.

Hour Circle
Any great circle on the celestial sphere whose plane is perpendicular to the plane of the celestial equator.

Index Error
A systematic error caused by deviation of an index mark or zero mark on an instrument having a scale or vernier, so that the instrument gives a non-zero reading when it should give a reading of zero. The distance error from the foot of a leveling rod to the nominal origin (theoretical zero) of the scale.

Indirect Leveling
The determination of differences of elevation from vertical angles and horizontal distances.

Interior Angle
An angle between adjacent sides of a closed figure and lying on the inside of the figure. The three angles within a triangle are interior angles.

International Foot
Defined by the ratio 30.48/100 meters.

International System of Units (SI)
A self-consistent system of units adopted by the general Conference on Weights and Measures in 1960 as a modification of the then-existing metric system.

Interpolation Method
Determination of a intermediate value between given values using a known or assumed rate of change of the values between the given values.

Intersection
Determining the horizontal position of a point by observations from two or more points of known position. Thus measuring directions or distances that intersect at the station being located. A station whose horizontal position is located by intersection is known as an intersection station.

Intervisibility
When two stations are visible to each other in a survey net.

Invar
An alloy of iron containing nickel, and small amounts of chromium to increase hardness, manganese to facilitate drawing, and carbon to raise the elastic limit, and having a very low coefficient of thermal expansion (about 1/25 that of steel).

Isogonic Chart
A system of isogonic lines, each for a different value of the magnetic declination.

Isogonic Line
A line drawn on a chart or map and connecting all points representing points on the earth having equal magnetic declination at a given time.

Laplace Azimuth
A geodetic azimuth derived from an astronomic azimuth by use of the Laplace equation.

Laplace Condition
Arises from the fact that a deflection of the vertical in the plane of the prime vertical will give a difference between astronomic and geodetic longitude and between astronomic and geodetic azimuth. Conversely, the observed differences between astronomic and geodetic values of the longitude and of the azimuth may both be used to determine the deflection in the plane of the prime vertical.

Laplace Equation
Expresses the relationship between astronomic and geodetic azimuths in terms of astronomic and geodetic longitudes and geodetic latitude.

Laplace Station
A triangulation or traverse station at which a Laplace azimuth is determined. At a Laplace station both astronomic longitude and astronomic azimuth are determined.

Least Count
The finest reading that can be made directly (without estimation) from a vernier or micrometer.

Least Squares Adjustment
The adjustment of the values of either the measured angles or the measured distances in a traverse using the condition that the sum of the squares of the residuals is a minimum.

Length of Closure
The distance defined by the equation:
$[(\text{closure of latitude})^2 + (\text{closure of departure})^2]^{0.5}$

Level
Any device sensitive to the direction of gravity and used to indicate directions perpendicular to that of gravity at a point.

Level Datum
A level surface to which elevations are referred. The generally adopted level datum for leveling in the US is mean sea level. For local surveys, an arbitrary level datum is often adopted and defined in terms of an assumed elevation for some physical mark.

Level Net
Lines of spirit leveling connected together to form a system of loops or circuits extending over an area.

Line of Sight
The line extending from an instrument along which distant objects are seen, when viewed with a telescope or other sighting device.

Local Coordinate System
Where the coordinate system origin is assigned arbitrary values and is within the region being surveyed and used principally for points within that region.

Local Datum
Defines a coordinate system that is used only over a region of very limited extent.

Loop Traverse
A closed traverse that starts and ends at the same station. A pattern of measurements in the field, so that the final measurement is made at the same place as the first measurement.

Magnetic Bearing
The angle with respect to magnetic north or magnetic south stated as east or west of the magnetic meridian.

Magnetic Meridian
The vertical plane through the magnetic pole including the direction, at any point, of the horizontal component of the Earth's magnetic field.

Major Semi-Axis
The line from the center of an ellipse to the extremity of the longest diameter. The term is also used to mean the length of the line.

Map
A conventional representation, usually on a plane surface and at an established scale, of the physical features (natural, artificial, or both) of a part or whole of the Earth's surface by means of signs and symbols and with the means of orientation indicated.

Map Accuracy
The accuracy with which a map represents. Three types of error commonly occur on maps: errors of representation, which occur because conventional signs must be used to represent natural or man-made features such as forests, buildings and cities; errors of identification, which occur because a non-existent feature is shown or is misidentified; and errors of position, which occur when an object is shown in the wrong position. Errors of position are commonly classified into two types: errors of horizontal location and errors of elevation. A third type, often neglected, is errors of orientation.

Map Scale
The ratio of a specified distance on a map to the corresponding distance in the mapped object.

Mean Angle
Average value of the angles.

Mean Lower Low Water (MLLW)
The average height of all lower low waters recorded over a 19-year period.
Mean Sea Level Datum
Adopted as a standard datum for heights or elevations. The Sea Level Datum of 1929, the current standard for geodetic leveling in the United States, is based on tidal observations over a number of years at various tide stations along the coasts.

Metric Unit
Belonging to or derived from the SI system of units.

Micrometer
In general, any instrument for measuring small distances very accurately. In astronomy and geodesy, a device, for attachment to a telescope or microscope, consisting of a mark moved across the field of view by a screw connected to a graduated drum and vernier. If the mark is a hair-like filament, the micrometer is called a filar micrometer.

Minor Semi-Axis
The line from the center of an ellipse to the extremity of the shortest diameter. I.e., one of the two shortest lines from the center to the ellipse. The term is also used to mean the length of the line.

Misclosure
The difference between a computed and measured value.

Monument
A physical object used as an indication of the position on the ground of a survey station.

NADCON
The National Geodetic Survey developed the conversion program NADCON (North American Datum Conversion) to convert to and from North American Datum of 1983. The technique used is based on a bi-harmonic equation classically used to model plate deflections. NADCON works exclusively in geographical coordinates (latitude/longitude).

Nadir
The point directly beneath the instrument and directly opposite to the zenith or the lowest point.

National Geodetic Vertical Datum 1929
Formerly adopted as the standard geodetic datum for heights, based on an adjustment holding 26 primary tide stations in North America fixed.

National Map Accuracy Standards
Specifications of the accuracy required of topographic maps published by the US at various scales.

National Tidal Datum Epoch
A period of 19 years adopted by the National Ocean Survey as the period over which observations of tides are to be taken and reduced to average values for tidal datums.

Network
Interconnected system of surveyed points.

Non-SI units
Units of measurement not associated with International System of Units (SI).

North American Datum of 1927
Formerly adopted as the standard geodetic datum for horizontal positioning. Based on the Clarke ellipsoid of 1866, the geodetic positions of this system are derived from a readjustment of survey observations throughout North America.

North American Datum of 1983
Adopted as the standard geodetic datum for horizontal positioning. Based on the Geodetic Reference System of 1980, the geodetic positions of this system are derived from a readjustment of survey observations throughout North America.

North American Vertical Datum of 1988
Adopted as the standard geodetic datum for heights.

Northing
A linear distance, in the coordinate system of a map grid, northwards from the east-west line through the origin (or false origin).

Open Traverse
Begins from a station of known or adopted position, but does not end upon such a station.

Optical Micrometer
Consists of a prism or lens placed in the path of light entering a telescope and rotatable, by means of a graduated linkage, about a horizontal axis perpendicular to the optical axis of the telescope axis. Also called an optical-mechanical compensator. The device is usually placed in front of the objective of a telescope, but may be placed immediately after it. The parallel-plate optical micrometer is the form usually found in leveling instruments.

Optical Plummet
A small telescope having a 90° bend in its optical axis and attached to an instrument in such a way that the line of sight proceeds horizontally from the eyepiece to a point on the vertical axis of the instrument and from that point vertically downwards. In use, the observer, looking into the plummet, brings a point on the instrument vertically above a specified point (usually a geodetic or other mark) below it.

Order of Accuracy
Defines the general accuracy of the measurements made in a survey. The order of accuracy of surveys are divided into four classes labeled: First Order, Second Order, Third Order and Fourth or lower order.

Origin
That point in a coordinate system which has defined initial coordinates and not coordinates determined by measurement. This point is usually given the coordinates (0,0) in a coordinate system in the plane and (0,0,0) in a coordinate system in space.

Orthometric Height
The elevation H of a point above or below the geoid.

Parallax
The apparent displacement of the position of a body, with respect to a reference point or system, caused by a shift in the point of observation.

Philadelphia Leveling Rod
Having a target but with graduations so styled that the rod may also be used as a self-reading leveling rod. Also called a Philadelphia rod. If a length greater than 7 feet is needed, the target is clamped at 7 feet and raised by extending the rod. When the target is used, the rod is read by vernier to 0.001 foot. When the rod is used as a self-reading leveling rod, the rod is read to 0.005 foot.

Photogrammetry
Deducing the physical dimensions of objects from measurements on photographs of the objects.

Picture Point
A terrain feature easily identified on an aerial photograph and whose horizontal or vertical position or both have been determined by survey measurements. Picture points are marked on the aerial photographs by the surveyor, and are used by the photomapper.

Planetable
A field device for plotting the lines of a survey directly from observations. It consists essentially of a drawing board mounted on a tripod, with a leveling device designed as part of the board and tripod.

Planimetric Feature
Item detailed on a planimetric map.

Plumb Line
The direction normal to the geopotential field. The continuous curve to which the gradient of gravity is everywhere tangential.

Positional Error
The amount by which the actual location of a cartographic feature fails to agree with the feature's true position.

Precision
The amount by which a measurement deviates from its mean.

Prime Meridian
The meridian of longitude 0°, used as the origin for measurement of longitude. The meridian of Greenwich, England, is almost universally used for this purpose.

Prime Vertical
The vertical circle through the east and west points of the horizon. It may be true, magnetic, compass or grid depending upon which east or west points are involved.

Project Control
Control used for a specific project.

Project Datum
Datum used for a specific project.

Projection
A set of functions, or the corresponding geometric constructions, relating points on one surface to points on another surface. A projection requires every point on the first surface to correspond one-to-one to points on the second surface.

Quadrangle
Consisting of four specified points and the lines or line segments on which they lie. The quadrangle and the quadrilateral differ in that the quadrangle is defined by four specified angle points, the quadrilateral by four specified lines or line-segments.

Random Error
Randomly distributed deviations from the mean value.

Range Pole
A simple rod fitted with a sharp-pointed, shoe of steel and usually painted alternately in red and white bands at 1-foot intervals.

Readings
The observed value obtained by noting and/or recording scales.

Real-time
An event or measurement reported or recorded at the same time as the event is occurring through the absence of delay in getting, sending and receiving data.

Reciprocal Leveling
Measuring vertical angles or making rod readings from two instrument positions for the purpose of compensating for the effects of refraction.

Rectangular Coordinate Systems
Coordinates on any system in which the axes of reference intersect at right angles.

Redundant Measurements
Taking more measurements than are minimally required for a unique solution.

Reference Meridian, True
Based on the astronomical meridian.

Reference Meridian, Magnetic
Based on the magnetic pole.

Reference Point
Used as an origin from which measurements are taken or to which measurements are referred.

Rejection Criterion
Probabilistic confidence limit used to compare with measurements to determine if the measurements are behaving according to a hypothesized prediction.

Refraction
The bending of rays by the substance through which the rays pass. The amount and direction of bending are determined by its refractive index.

Relative Accuracy
Indicated by the dimensions of the relative confidence ellipse between two points. A quantity expressing the effect of random errors on the location of one point or feature with respect to another.

Repeating Theodolite
Designed so that the sum of successive measurements of an angle can be read directly on the graduated horizontal circle.

Resection
Determining the location of a point by extending lines of known direction to two other known points.

Sexagesimal System
Notation by increments of 60. As the division of the circle into 360°, each degree into 60 minutes, and each minute into 60 seconds.

Set-up
In general, the situation in which a surveying instrument is in position at a point from which observations are made.

Spheroid
Used as a synonym for ellipsoid.

Spirit Level
A closed glass tube (vial) of circular cross section. Its center line forms a circular arc with precise form and filled with ether or liquid of low viscosity, with enough free space left for a bubble of air or gas.

Stadia Constant
The sum of the focal length of a telescope and the distance from the vertical axis of the instrument on which the telescope is mounted to the center of the objective lens-system.

Stadia Traverse
Distances are determined using a stadia rod. A stadia traverse is suited to regions of moderate relief with an adequate network of roads. If done carefully, such a traverse can establish elevations accurate enough for compiling maps with any contour interval now standard.

Standard Error
The standard deviation of the errors associated with physical measurements of an unknown quantity, or statistical estimates of an unknown quantity or of a random variable.

Systematic Error
Errors that affect the position (bias) of the mean. Systematic errors are due to unmodeled affects on the measurements that have a constant or systematic value.

State Plane Coordinate System (SPCS)
A planar reference coordinate system used in the United States.

Strength of Figure
A number relating the precision in positioning with the geometry with which measurements are made.

Subtense Bar
A bar with two marks at a fixed, known distance apart used for determining the horizontal distance from an observer by means of the measuring the angle subtended at the observer between the marks.

Taping
Measuring a distance on the using a surveyor's tape.
Three-wire Leveling
The scale on the leveling rod is read at each of the three lines and the average is used for the final result.

Topographic Map
A map showing the horizontal and vertical locations of the natural and man-made features represented and the projected elevations of the surroundings.

Transformation
Converting a position from one coordinate system to another.

Transit
The apparent passage of a star or other celestial body across a defined line of the celestial sphere.

Transit Rule
The correction to be applied to the departure (or latitude) of any course has the same ratio to the total misclosure in departure (or latitude) as the departure (latitude) of the course has to the arithmetical sum of all the departures (latitudes) in the traverse. The transit rule is often used when it is believed that the misclosure is caused less by errors in the measured angles than by errors in the measured distances.

Transverse Mercator Projection
Mercator map projection calculated for a cylinder with axis in the equatorial plane.

Traverse
A sequence of points along which surveying measurements are made.

Triangulation
Determination of positions in a network by the measurement of angles between stations.

tribrach
The three-armed base, of a surveying instrument, in which the foot screws used in leveling the instrument are placed at the ends of the arms. Also called a leveling base or leveling head.

Trigonometric heighting
The trigonometric determination of differences of elevation from observed vertical angles and measured distances.

Trilateration
Determination of positions in a network by the measurement of distances between stations using the intersection of two or more distances to a point.

Universal Transverse Mercator
A worldwide metric military coordinate system.

US Coast & Geodetic Survey (USC&GS)
Now known as National Ocean Service (NOS).

US Survey Foot
The unit of length defined by 1200/3937 m

Variance-Covariance Matrix
A matrix whose elements along the main diagonal are called the variances of the corresponding variables; the elements off the main diagonal are called the covariances.

Vernier
An auxiliary scale used in reading a primary scale. The total length of a given number of divisions on a vernier is equal to the total length of one more or one less than the same number of divisions on the primary scaled.

VERTCON
Acronym for vertical datum conversion. VERTCON is the computer software that converts orthometric heights between NGVD 29 to NAVD 88.

Vertical Angle
An angle in a vertical plane either in elevation or depression from the horizontal.

Vertical Circle
A graduated scale mounted on an instrument used to measure vertical angles.

Vertical Datum
Any level surface used as a reference for elevations. Although a level surface is not a plane, the vertical datum is frequently referred to as the datum plane.

World Geodetic System of 1984
Adopted as the standard geodetic datum for GPS positioning. Based on the World Geodetic System reference ellipsoid.

Wye Level
Having the telescope and attached spirit level supported in wyes (Y's) in which it can be rotated about its longitudinal axis (collimation axis) and from which it can be lifted and reversed, end for end. Also called a Y-level and wye-type leveling instrument.

Zenith
The point above the instrument where an extension of a plumb (vertical) line at the observer's position intersects the celestial sphere.

Zenith Angle
Measured in a positive direction downwards from the observer's zenith to the observed target.

Zenith Distance
The complement of the altitude, the angular distance from the zenith of the celestial body measured along a vertical circle.